Lecture Notes in Computer S

T0238246

Commenced Publication in 1973
Founding and Former Series Editors:
Gerhard Goos, Juris Hartmanis, and Jan van Leeuwen

Giuseppe Psaila   Roland Wagner (Eds.)

E-Commerce and Web Technologies

8th International Conference, EC-Web 2007
Regensburg, Germany, September 3-7, 2007
Proceedings

 Springer

Volume Editors

Giuseppe Psaila
Università degli studi di Bergamo
Facoltà di Ingegneria, Viale Marconi 5, 24044 Sede di Dalmine (BG), Italy
E-mail: psaila@unibg.it

Roland Wagner
University of Linz
Institute of FAW
Altenbergerstrasse 69, 4040 Linz, Austria
E-mail: rrwagner@faw.uni-linz.ac.at

Library of Congress Control Number: 2007933402

CR Subject Classification (1998): H.4, K.4.4, J.1, K.5, H.3, H.2, H.2.5, K.6.5

LNCS Sublibrary: SL 3 – Information Systems and Application, incl. Internet/Web
and HCI

ISSN   0302-9743
ISBN-10   3-540-74562-9 Springer Berlin Heidelberg New York
ISBN-13   978-3-540-74562-4 Springer Berlin Heidelberg New York

Springer is a part of Springer Science+Business Media

springer.com

© Springer-Verlag Berlin Heidelberg 2007
Printed in Germany

Typesetting: Camera-ready by author, data conversion by Scientific Publishing Services, Chennai, India
Printed on acid-free paper   SPIN: 12115075   06/3180   5 4 3 2 1 0

Preface

Welcome to EC-Web 2007, the International Conference on E-commerce and Web Technologies. As in the past 7 years, EC-Web was co-located with DEXA, the International Conference on Database and Expert Systems Applications, and took place in Regensburg, a very beautiful city in the heart of Europe.

This was the eighth edition of EC-Web, and we think that it is now a mature conference. It gained a stable audience and it is considered by several authors as a key conference for publishing their ideas and their work.

The new Chairs, Roland Wagner and Giuseppe Psaila, followed the line of evolution to make the conference even more attractive. In fact, one key feature of EC-Web is the two-fold nature of the conference: it brings together both papers proposing technological solutions for E-commerce and the World Wide Web, and papers concerning management of E-commerce, such as Web marketing, impact of E-commerce on business processes and organizations, analysis of case studies. This year, the new topic of "social aspects" was introduced: in fact, the every-day increasing availability of E-commerce solutions for consumers is causing the rise of new behaviors that must be studied, in order to understand the impact of E-commerce solutions on every-day life and the new opportunities that these new behaviors open.

The second significant change introduced this year was the number of reviewers: the Program Committee was composed of more than 120 reviewers (instead of 60 reviewers in the last edition). This choice was motivated by the wish to provide a better service to authors and improve the quality of the selection process.

The technical program, comprised 22 papers selected among 67 submitted papers, with a selection rate of 32%. The contributions covered several interesting areas, such as security and privacy, Web services, recommender systems, Web marketing, profiling and customer behavior, electronic commerce technology, impact of E-commerce on organizations.

We think the program is interesting and we hope the readers think the same.

June 2007

Giuseppe Psaila
Roland R. Wagner

Organization

Program Committee Chairpersons

Giuseppe Psaila, University of Bergamo, Italy
Roland Wagner, FAW, University of Linz, Austria

Program Committee

Marco Aiello, Rijksuniversiteit Groningen, The Netherlands
Esma Aïmeur, University of Montreal, Canada
Damminda Alahakoon, Monash University, Australia
Sergio Alonso, University of Granada, Spain
Jörn Altmann, Seoul National University, South-Korea and Intl. University of
    Bruchsal, Germany
Sami Bhiri, DERI, Ireland
Sourav S. Bhowmick, Nanyang Technological University, Singapore
Enrique Bigné, University of Valencia, Spain
Susanne Boll, University of Oldenburg, Germany
Michelangelo Ceci, University of Basi, Italy
Wojciech Cellary, Poznan University of Economics, Poland
Francisco Chiclana, De Montfort University, UK
Jen-Yao Chung, IBM T.J. Watson Research Center, USA
Andrzej Cichocki, Telcordia Technologies, USA
Kajal Claypool, Oracle, USA
Emmanuel Coquery, University Lyon 1, France
Arthur I. Csetenyi, Budapest Corvinus University, Hungary
Alfredo Cuzzocrea, University of Calabria, Italy
Simão Melo de Sousa, University of Beira Interior, Portugal
Radoslav Delina, Technical University of Kosice, Slovakia
Tommaso Di Noia, Politecnico di Bari, Italy
Schahram Dustdar, Vienna University of Technology, Austria
Johann Eder, University of Vienna, Austria
Maria Jose Escalona, Universidad de Sevilla, Spain
Torsten Eymann, University of Bayreuth, Germany
Eduardo Fernandez, Florida Atlantic University, USA
Gianluigi Ferrari, University of Pisa, Italy
Elena Ferrari, University of Insubria at Como, Italy
Ludger Fiege, Siemens, Germany
Carlos Flavian, University of Zaragoza, Spain
Farshad Fotouhi, Wayne State University, USA
Eduard Cristóbal Fransi, University of Lleida, Spain
Yongjian Fu, Cleveland State University, USA

Walid Gaaloul, DERI, Ireland
Stephane Gagnon, Université du Québec en Outaouais (UQO), Canada
Jing Gao, University of South Australia, Australia
Peter Geczy, AIST, Japan
Chanan Glezer, Ben Gurion University, Israel
Claude Godart, University of Nancy and INRIA, France
Mohand-Said Hacid, University Lyon 1, France
G. Harindranath, University of London, UK
Aboul Ella Hassanien, Kuwait University, Kuwait
Josef Herget, University of Chur, Switzerland
Enrique Herrera-Viedma, University of Granada, Spain
Klaus Herrmann, University of Stuttgart, Germany
Charles Hofacker, Florida State University, USA
Yigal Hoffner, IBM Zurich Research Lab., Switzerland
Birgit Hofreiter, University of Vienna, Austria
Christian Huemer, Vienna University of Technology, Austria
Michael C. Jaeger, Berlin University of Technology, Germany
Dimka Karastoyanova, University of Stuttgart, Germany
Gregory E. Kersten, Concordia University Montreal, Canada
Hiroyuki Kitagawa, University of Tsukuba, Japan
Jan Klas, University of Economics, Prague, Czech Republic
Gabriele Kotsis, Johannes Kepler University Linz, Austria
Sandeep Krishnamurthy, University of Washington, USA
Alberto Laender, Federal University of Minas Gerais, Brazil
Deok Gyu Lee, Electronics and Telecommunications Research Institute(ETRI), Korea
Juhnyoung Lee, IBM T.J. Watson Research Center, USA
Joerg Leukel, University of Hohenheim, Germany
Leszek T. Lilien, Western Michigan University, USA
Ee-Peng Lim, Nanyang Technological University, Singapore
Huan Liu, Arizona State University, USA
Antonio Gabriel Lopez, University of Granada, Spain
Heiko Ludwig, IBM T.J. Watson Research Center, USA
Sanjay Kumar Madria, University of Missouri-Rolla, USA
Koné Mamadou Tadiou, Université Laval, Canada
Mário Marques Freire, University of Beira Interior, Portugal
Luis Martínez Lopez, University of Jaen, Spain
Francisco Mata Mata, University of Jaen, Spain
Massimo Mecella, University of Rome La Sapienza, Italy
Bamshad Mobasher, DePaul University, USA
Mukesh Mohania, IBM India Research Lab, India
Gero Muehl, TU Berlin, Germany
Guenter Mueller, University of Freiburg, Germany
Dirk Neumann, University of Karlsruhe, Germany
Wee-Keong Ng, Nanyang Tech. University, Singapore
Anne-Marie Oostveen, Oxford Internet Institute, UK
Rolf Oppliger, eSECURITY Technologies, Switzerland
Stefano Paraboschi, University of Bergamo, Italy

Acknowledgement

The work was supported by the PRIN 2006 program of the Italian Ministry of Research, within project "Basi di dati crittografate" (2006099978).

Table of Contents

Web Services

E-Commerece and Organizations

Web Marketing

EC Technology

A Secure Payment Protocol for Restricted Connectivity Scenarios in M-Commerce

Jesús Téllez Isaac[1] and José Sierra Camara[2]

[1] Universidad de Carabobo, Computer Science Department (Facyt)
Av. Universidad, Sector Bárbula, Valencia, Venezuela
jtellez@uc.edu.ve
[2] Universidad Carlos III de Madrid, Computer Science Department,
Avda. de la Universidad, 30, 28911, Leganés (Madrid), Spain
sierra@inf.uc3m.es

Abstract. A significant number of mobile payment systems have been proposed in recent years, most of them based on a scenario where all the entities are directly connected one to another (formally called "Full connectivity scenario"). Despite of the advantages that the aforementioned scenario offers to protocol's designers, regarding design simplification and development of payment protocols without losing security capabilities, the full connectivity scenario does not consider those situations in which the client cannot directly communicate with the issuer (Kiosk Centric Model) or the merchant has no direct communication with the acquirer (Client Centric Model). In order to overcome this restriction and contribute to the progress of m-commerce, in this paper we propose an anonymous protocol that uses a digital signature scheme with message recovery using self-certified public keys that is suitable for both the Kiosk Centric Model and Client Centric Model. As a result, our proposal shows that m-commerce is possible in restrictive connectivity scenarios, achieving the same security capabilities than other protocols designed for mobile payment systems based on "Full connectivity scenario".

Keywords: Payment Protocol, Self-certified public keys, Digital Signature with message recovery, Mobile Payment System.

1 Introduction

Several mobile payment systems have emerged in the last years which allow payments for services and goods from mobile devices using different kinds of payments: credit-card payments, micropayments and digital coins. The relationship between payee and acquirer is quite strict in most of these mobile payment systems and does not allow the use of schemes in which the communication among these parties is not possible due to: 1) the impossibility of the merchant to connect to Internet and 2) the high costs and/or inconveniences of using the infrastructure necessary to implement other mechanisms of communication between the merchant and the acquirer (such SMS, phone call, etc.).

G. Psaila and R. Wagner (Eds.): EC-Web 2007, LNCS 4655, pp. 1–10, 2007.

The above restrictions do not represent an important issue for the majority of mobile payment systems proposed up until now because they assume that engaging parties are able to connect to Internet. Nevertheless, in the real world there are some situations that the merchant meets in which it is not possible to connect to the Internet, so it becomes necessary to develop mobile payment systems where the payee could sell goods/services even thought he/she may not have Internet access.

According to our operational models (where client cannot communicate directly with issuer, or merchant cannot communicate with the acquirer in a direct way, the traditional digital signature schemes based on asymmetric techniques are not suitable because one party (client or merchant, depending on the scenario) has connectivity restrictions and consequently, communication with others parties (as a CA, for verifying a certificate) is not possible during a purchase. Therefore, usage of a non-traditional digital signature scheme is required in order to satisfy our requirements.

In order to eliminate the restriction of those mobile payment systems based on the Full Connectivity Scenario regarding the direct communication between client and issuer, and among merchant and acquirer for authentication purposes, in section 3, we design a protocol that allows to a party (A) to send a message to another peer (B) through a third party (who will not be able to decrypt this message) in the those scenarios. The proposed protocol employs the authentication encryption scheme proposed by [13] that allows only specified receivers to verify and recover the message, so any other receiver will not be able to access the information. Moreover, it supports both credit-card and debit-card transactions and protects the real identity of the clients during the purchase. As a result, our proposal represents an alternative to other mobile payment systems with restrictions regarding a mandatory connection among two of its parties.

Outline of this paper: We begin by presenting the related work. Then, we present our approach which includes a complete list of notations, the operational model and the proposed protocol. In section 4, a security analysis of the proposed protocol is presented. We end this paper with the conclusions in section 5.

2   Related Work

Recently, [3] conducted a research that unifies many proposed m–commerce usages into a single framework. This research intended to revise the possible range of mobility scenarios, identifying the security issues for each connectivity scenario. As a result, five scenarios were identified and analyzed: Disconnected Interaction, Server Centric Case, Full Connectivity, Kiosk Centric Case and Client Centric Case. The last two have been considered as the starting point in the design of our proposal.

Most of the protocols proposed in recent years for the Full Connectivity scenario are based on public-key infrastructure (PKI) [1,4,8,12] whereas the remaining employ symmetric-key operations which is more suitable for wireless networks [7]. Unfortunately, usage of those protocols is not possible in scenarios where direct interaction among two of its parties is not allowed due to the communication restriction imposed by the model (as happens in Kiosk Centric Model or Client Centric Model). However, some protocols could be reformulated to overcome this restriction (achieving the same security and performance levels, but in a different scenario), while being suitable for mobile payment systems with Restricted Connectivity. For example, Téllez *et al.* [9] reformulate the mobile payment protocol proposed by [7] to satisfy the requirements of their proposal.

A few number of signatures schemes with message recovery have been proposed in recent years which illustrate how a signer's public key can be simultaneously securely authenticated during the signature verification, avoiding communication with a Certificate Authority during a transaction in order to verify the validity of a certificate since the certificate is embedded in public key itself. Therefore, and as shown in [10], digital scheme signature schemes with message recovery are suitable for mobile payment protocols based on a restrictive connectivity scenarios like the one being suggested in this work.

3 Our Approach

3.1 Parties and Notations

All the entities involved in our protocol are called parties and communicate through wireless and wired network. The symbols C, M, PG, I, A are used to denote the names of the parties Client, Merchant, Payment Gateway, Issuer and Acquirer respectively. The following symbols are used to represent other messages and protocols:

- ID_P : the identity of party P that contains the contact information of P.
- NID_C : Client's nickname, temporary identity.
- K_P : party's K public key.
- K_S : party's K private key.
- $E_{P-P'}(X)$: message X signed and encrypted by ID_P to a specified receiver $ID_{P'}$, following the generation procedure of signature proposed by [13].
- TID: Identity of transaction that includes time and date of the transaction.
- OI: Order information (OI = {TID, OD, h(OD, Price)}) where OD and Price are order descriptions and its amount.
- TC: The type of card used in the purchase process (TC={Credit, Debit}).
- TS: The type of scenario used during a payment (TC={Kiosk, Client})
- DCMA : The status of the direct connection between the merchant and the acquirer (DCMA = {Connected, NO-Connected}). The default value is *NO-Connected*.
- DCCI : The status of the direct connection between the client and the issuer (DCCI = {Connected, NO-Connected}). The default value is *NO-Connected*.
- Stt: The status of transaction (Stt = {Accepted, Rejected}).
- TIDReq : The request for TID.
- MIDReq : The request for ID_M.
- MPReq : The request for M_P.
- DCMAReq : The request for DCMA.
- $h(M)$: the one-way hash function of the message M.

3.2 Operational Model

Our operational models, Kiosk Centric Model and Client Centric Model (figure 1 and figure 2, respectively), are composed of five entities:

1. *Client*: a user who wants to buy goods or services from the merchant, equipped with a short range link (such Infrared, Wi-Fi or Bluetooth). Only in the Client Centric Model, the client is able to access Internet.
2. *Merchant*: a computational entity (such an intelligent vending machine) that offers or sells products or services to the client, and with which the user participates in a transaction using a short range link. In Kiosk Centric Model, this entity connects with the Payment Gateway through a secure channel allowing the merchant to communicate with the acquirer using this connection whereas in Client Centric Model, direct communication with the Issuer is not possible so it must take place through the client.
3. *Acquirer*: is the merchant's financial institution.
4. *Issuer*: is the customer's financial institution.
5. *Payment Gateway*: an additional entity that acts as a medium between acquirer/issuer at banking private network side and client/vendor at the Internet side for clearing purpose [7].

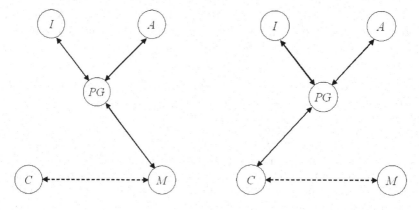

Fig. 1. Kiosk Centric Model   **Fig. 2.** Client Centric Model

The links among the five entities of our operational models are specified in figure 1 and 2. Note that, in both operational models, the connection between the client and the merchant (denoted as the dotted arrow) is setup up through a wireless channel.

On the other hand, interaction among client and payment gateway or between merchant and payment gateway (depicted as the solid arrow in any of the operational models) should be reliable and secure against passive and active attacks. Note that the issuer, acquirer and payment gateway operates under the banking private network, so the security of the messages exchanged among them is out of the scope of this paper.

3.3 Initial Assumptions

The initial assumptions for our proposed protocol can be stated as follows:

1. Client registers herself to an issuer before making payments. The registration can be done either personally at the issuer's premises or via the issuer's website. During the above process, the client shares her credit- and/or debit-card information (CDCI) with her issuer (who will not reveal it to any merchant). On the other hand, the issuer assigns several nicknames to the client and those nicknames are know only by the client and the issuer [6]. In the Kiosk Centric Model, the client sends (with the assistance of the issuer) her nicknames and x_C to SA and receives all system parameters from the SA.
2. The system authority (SA) is responsible for generation of the system parameters in the system initialization phase (as described in [13][11]).
3. Every party of the system P_i (whose identity is ID_{P_i}) choose a number K_{S_i} as her secret key and computes $x_i = g^{K_{S_i}} \bmod N$. Then, P_i sends (x_i, ID_{P_i}) to SA. After receiving (x_i, ID_{P_i}), the SA computes and publishes the public key of P_i as $K_{P_i} = (x_i - ID_{P_i})^{h^{-1}(ID_{P_i})} \bmod N$ [13]. As the client uses a nickname instead the real identity to protect her privacy, one K_{P_i} must be generated and published for every nickname assigned to the client.
4. The client holds C_S, ID_I, and system parameters in her mobile device. Also, in Kiosk Centric Model, client holds I_P.

3.4 Detailed Protocols

Our Protocol consists of two sub–protocols: the *Merchant Registration Protocol* and the *Payment Protocol*. The main functions of both protocols are shown as follows:

Merchant Registration Protocol

> **C → M:** $\{NID_C, n, MIDReq, DCMAReq\}_w$
> **M → C:** $\{ID_M, h(n, NID_C, ID_M)\}_w$
> **C → M:** IF ($TS = "Kiosk"$) THEN
> $\qquad \{n, MPReq\}_w$
> ELSE
> $\qquad \{n, C_P\}_w$
> **M → C:** IF ($TS = "Kiosk"$) THEN
> $\qquad \{n, M_P, h(n, M_P)\}_w$
> ELSE
> $\qquad \{n, C_P, h(n, C_P)\}_w$

As our protocol is designed to work on two different operational models, the first step is to determine in which one them the payment is going to take place. First, **C** assigns the value *Connected* to $DCCI$ if he/she is able to connect to internet from his/her mobile device. Then, **C** sends to **M** her nickname NID_C, a nonce n for challenge-response, $MIDReq$, $DCMAReq$ and $h(n, NID_C, ID_M)$, encrypted with a session key w generated by running AKE protocol [1] with C. After $\mathbf{M}$ receives the message, he/she sends the merchant's identity, $DCMA$, encrypted with the session key w. Note that $\mathbf{M}$ assigns the value *Connected* is she/he has direct communication with the acquirer.

After $\mathbf{C}$ receives $DCMA$, he/she will determine the operational model to be used by the protocol, comparing $DCCI$ with $DCMA$ and takes the decision following this rule:

> IF ($DCCI$="NO-Connected") AND ($DCMA$ = "Connected") THEN
>   ASSIGN the value "*Kiosk*" TO TS
> ELSE
>   ASSIGN the value "*Client*" TO TS

Once the operational model to be used has been decided, $\mathbf{C}$ prepares a new message and sends it to $\mathbf{M}$. The message includes a new nonce n for challenge-response and MPReq (if the value of TS equals "*Kiosk*") or C_P (when the value of TS is "*Client*"). After $\mathbf{M}$ receives the message, $\mathbf{M}$ confirms C's registration by sending another message, encrypted with the session key w. This message includes n, M_P and $h(n, M_P)$ when TS has the value "Client", or n, C_P and $h(n, C_P)$ if the value of TS equals to "Kiosk".

Afterwards, if the wallet software is not available in the mobile device, the client will obtain it through one of the following methods: 1) Sending a software request to I, in the Kiosk Centric Model, or 2) Connecting to issuer's web site to download it or sending a request to the issuer to receive it by mail. This method is valid for the Client Centric Model.

Once the client has received the software, she will install it in the mobile device. Note that, in the Kiosk Centric Model, the software is signed using the Authenticated encryption scheme with message linkages proposed by [2,13].

Payment Protocol

1) $\mathbf{C} \rightarrow \mathbf{M}$: $NID_C, TIDReq$
   $\mathbf{M} \rightarrow \mathbf{C}$: $E_{M-C}(TID, ID_M)$

2) $\mathbf{C} \rightarrow \mathbf{M}$: $E_{C-M}(OI, Price, NID_C, ID_I, VSRequest)$
   $VSRequest = E_{C-I}(Price, OI, TC, ID_M)$

3) **Merchant:** IF (TS = "*Kiosk*") THEN
   $\mathbf{M} \rightarrow \mathbf{PG}$: $VCRequest, ID_M, TS$
   ELSE
   $\mathbf{M} \rightarrow \mathbf{C}$: $E_{M-C}(VCRequest)$
   $VCRequest = E_{M-PG}(VSRequest, h(OI), TID, Price, NID_C, ID_I)$

4) **Client:**   IF (TS = "*Client*") THEN
   $\mathbf{C} \rightarrow \mathbf{PG}$: $E_{C-PG}(VCRequest, ID_M, TS), NID_C$

5) Under banking private network,

    5.1) PG → I: $VSRequest, h(OI), TID, Price, NID_C, ID_M$

    5.2) PG → A: $Price, ID_M$

    5.3) I,A → PG: $VSResponse, Stt, h(Stt, h(OI))$

              $VSResponse = E_{I-C}(Stt, h(OI))$

6) Payment Gateway: IF $(TS = "Kiosk")$ THEN

                    PG → M: $VCResponse$

              ELSE

                    PG → C: $PResponse$

    $VCResponse = E_{PG-M}(Stt, VSResponse, h(Stt, h(OI)))$

    $PResponse\ = E_{PG-C}(VSResponse, VCResponse)$

7) Merchant: IF $(TS = "Kiosk")$ THEN

              M → PG: $E_{M-C}(PResponse)$

    $PResponse = E_{M-C}(VSResponse)$

8) Client: IF $(TS = "Client")$ THEN

              C → M: $E_{C-M}(PResponse)$

Step 1: The client **C** and the merchant **M** exchange the information necessary to start the protocol.

Step 2: **C** creates a *Payment Request* (referred to the General Payment Model described in [7]) including C's nickname, I's identity, *Price*, *OI* (used to inform **M** about the goods and prices requested) and *Value-Substraction Request* (called *VSRequest*, which is encrypted to be recovered only by an issuer **I** and includes *Price, TC, ID_M* and OI). The client **C** encrypts the *Payment Request* to be recovered only by **M** and sends it to the merchant.

Step 3: **M** decrypts the message received from **C** to retrieve *OI*. **M** prepares the *Value-Claim Request* (called *VCRequest*, which contains C's nickname, ID_I, *Price*, identity of transaction and order information), encrypted to be recovered only by **PG** in order to ensure that only the payment gateway is the intended recipient of the message. Once the *VCRequest* has been prepared, if the value of TS equals "*Kiosk*", **M** sends it to **PG** with M's identity and TS. Otherwise, **M** sends *VCRequest* to **C**, encrypted to be recovered only by the client.

Step 4: This step occurs only when the value of TS is "*Client*". **C** decrypts the message received from **M** to recover *VCRequest*. Then, **C** prepares a new message (which includes ID_M, TS and the forwarded $Value - Claim\ Request$) encrypted to the specific receiver **PG**. The later encrypted message is sent by the client **C** to **PG** with her/his nickname (NID_C).

Step 5: **PG** decrypts the message received from **C** to retrieve *VSRequest* and the others fields included in *VCRequest*. Then, **PG** forwards *VSRequest* and other important information, namely: *h(OI), TID, Price, NID_C, ID_I* to **I** who will process it to approve or reject the transaction. Also, **PG** sends ID_M and the requested price (*Price*) to claim to acquirer **A** that she is the party whom the requested amount *Price* will be transferred to. After checking the validity of the client's account, the total amount of *OI* is transferred to the merchant's account, the issuer **I** prepares *Value-Substraction Response* (called *VSResponse*) and sends it to **PG** with the approval result (*Stt*). Note that *VSResponse* is encrypted to be recovered only by an issuer **C**.

Step 6: If the value of TS equals "*Kiosk*", the payment gateway (**PG**) sends $Value - Claim\ Response$ (called *VCResponse*, that includes *VSResponse*, which will be forwarded to **C**) encrypted to be recovered only by **M**. Otherwise, **PG** sends *Payment Response* (called *PResponse*) encrypted to be recovered only by **C**. *PResponse* includes *VSResponse* and *VCResponse* (which will be forwarded to **M**).

Note that in the Kiosk Centric Model (when the value of TS is "*Kiosk*"), as **M** has her/his own *OI*, she/he can compare this field with the received *h(OI)* to check whether or not the message is the response of her/his request. If they are not matched, **M** sends a message to the **PG** pointing the problem.

Step 7: This step occurs only when the value of TS is "*Kiosk*". **M** encrypts $Value - Subtraction\ Response$ to be recovered only by **C**. Then, **M** sends it to **C** as *Payment Response* (called *PResponse*). Once **C** receives the message, decrypts it to retrieve the result of her/his request.

Step 8: This step is performed only if the value of TS equals "*Client*". After receiving *PResponse*, **C** decrypts it to retrieve *VSResponse* and *VCReponse*. Then, **C** compares her/his own *OI* with the received *h(OI)* to check whether or not the message is the response of her/his request. If they are not matched, **C** sends a message to the **PG** pointing the problem (then, the payment gateway can start a recovery procedure or resend the message). Otherwise, **C** encrypts $Value - Claim\ Response$ to be recovered only by **M** and sends it to the merchant who in turn proceeds to deliver the goods to the client.

Once the purchase has been completed, the client does not have to run *Merchant Registration Protocol* again unless she wants to perform transaction with a new merchant. Note that, after client finish all purchase with a merchant, she will remove merchant's information from her mobile device.

4   Analysis and Discussions

4.1   Security Issues

Transaction Security

- **Authentication:** Each one of the operational models used in this proposal have a communication restriction. Therefore, in order to allow a party (A) to authenticate another peer (B), a sender has to send a message to a specified receiver through a third party with the following features: 1) resistant to attacks while in transit, 2) recoverable only by the specified receiver, and 3) able to assure that it has been created and sent by the sender.

  As the authenticated encryption scheme used in our protocol integrates the mechanisms for signature and encryption, it satisfies all the requirements mentioned above and can be used by any party of the system to authenticate another peer in a secure way.

- **Confidentiality:** In our protocol, the important data of each transaction is protected while in transit because the scheme used enables only the specified receiver to verify & recover the message transmitted by the sender, while any other party cannot perform such operations.
- **Integrity:** Any important data should be protected from being modified or/and replaced while in transit. In our protocol, the integrity is ensured by the digital signature with message recovery technique.

Anonymity: In order to prevent a merchant from knowing the identity of a client, a nickname NID_C instead of his/her real identity is used during a communication from C to M. Therefore, as merchant cannot map the nickname and C'S true identity, client's privacy is protected and untraceable.

Non-Repudiation of Origin (NRO): As in our protocol, a signer U generates the signature of a message using his/her private key U_S (known only by U), he/she should not be able to repudiate his signature creation later. Therefore, Non-repudiation of Origin is ensured.

5 Conclusions and Further Work

We have proposed a secure protocol for secure payments in a mobile payment system where client cannot directly communicate with the issuer or the merchant has no direct communication with the acquirer. Therefore, the messages among the parties that can not communicate directly must be done across a third party.

The proposed protocol employs a digital signature scheme with message recovery using self-certified public keys which allows to a client to make purchases without disclosing private information and using a feasible connection with the merchant through a short range link.

Our proposal represents an alternative to all mobile payment systems based on the "Full Connectivity scenario" where communication among all the engaging parties is mandatory. As a result, we assert that our proposal shows how m-commerce is possible in restrictive connectivity scenarios, achieving the same security capabilities than other protocols designed for mobile payment systems based on "Full connectivity scenario".

As the digital scheme employed in the proposed protocol includes only non-repudiation of origin, it will be valuable in the future incorporate more non-repudiation services in order to prevent entities from denying that they have sent or received certain messages.

Acknowledgement

This work was supported in part by ASPECTS-M Project (Reference Models for Secure Architectures in Mobile Electronic Payments), CICYT-2004, however it represents the view of the authors.

References

1. Bellare, M., Garay, J., Hauser, R., Herzberg, A., Krawczyk, H., Steiner, M., Tsudik, G., Herreweghen, E., Waidner, M.: Design, implementation and deployment of the iKP secure electronic payment system. IEEE Journal on Selected Areas in Communications 18(4), 611–627 (2000)
2. Chang, Y., Chang, C., Huang, H.: Digital signature with message recovery using self-certified public keys without trustworthy system authority. Applied Mathematics and Computation 161(1), 211–227 (2005)
3. Chari, S., Kermani, P., Smith, S., Tassiulas, L.: Security issues in m-commerce: A usage based taxonomy. In: Liu, J., Ye, Y. (eds.) E-Commerce Agents. LNCS (LNAI), vol. 2033, pp. 264–282. Springer, Heidelberg (2001)
4. Hall, J., Kilbank, S., Barbeau, M., Kranakis, E.: WPP: A Secure Payment Protocol for Supporting Credit- and Debit-card Transactions Over Wireless Networks. In: IEEE International Conference on Telecommunications (ICT) (2001)
5. Ham, W., Choi, H., Xie, Y., Lee, M., Kim, K.: A secure one-way mobile payment system keeping low computation in mobile devices. In: WISA2002. LNCS, pp. 287–301. Springer, Heidelberg (2002)
6. Hu, Z., Liu, Y., Hu, X., Li, J.: Anonymous Micropayments Authentication (AMA) in Mobile Data Network. In: The 23rd Annual Joint Conference of the IEEE Computer and Communications Societies (IEEE INFOCOM), pp. 7–11 (2004)
7. Kungpisdan, S.: A secure account-based mobile payment system protocol. In: International Conference on Information Technology: Coding and Computing (ITCC), pp. 35–39 (2004)
8. Lei, Y., Chen, D., Jiang, Z.: Generating digital signatures on mobile devices. In: 18th International Conference on Advanced Information Networking and Applications (AINA 2004), pp. 532–535. IEEE Computer Society, Los Alamitos (2004)
9. Téllez, J., Sierra, J., Izquierdo, A., Carbonell, M.: Payment in a Kiosk Centric Model with Mobile and Low Computational Power Devices. In: Gavrilova, M., Gervasi, O., Kumar, V., Tan, C.J.K., Taniar, D., Laganà, A., Mun, Y., Choo, H. (eds.) ICCSA 2006. LNCS, vol. 3984, pp. 798–807. Springer, Heidelberg (2006)
10. Téllez, J., Sierra, J., Izquierdo, A., Márquez, J.: Anonymous Payment in a Kiosk Centric Model with Mobile using Digital signature scheme with message recovery and Low Computational Power Devices. Journal of Theoretical and Applied Electronic Commerce Research 1(2), 1–11 (2006)
11. Tseng, Y., Jan, J., Chien, H.: Digital signature with message recovery using self-certified public keys and its variants. Applied Mathematics and Computation 136(2-3), 203–214 (2003)
12. Wang, H., Kranakis, E.: Secure Wireless Payment Protocol. In: International Conference on Wireless Networks, pp. 576–578 (2003)
13. Zhang, J., Zou, W., Chen, D., Wang, Y.: On the Security of a Digital Signature with Message Recovery using Self-certified Public Key. Soft Computing in Multimedia Processing, Special Issue of the Informatica Journal 29(3), 343–346 (2005)

Using WPKI for Security of Web Transaction

Mohammed Assora, James Kadirire, and Ayoub Shirvani

Anglia Ruskin University
Chelmsford, Essex CM1 1LL, UK
{m.assora, j.kadirire, a.shirvani}@anglia.ac.uk

Abstract. Today, a web transaction is typically protected by using SSL/TLS. SSL/TLS without compulsion for a client's public key certificate, which is the typical usage, is not able to fulfill the security requirements for web transactions. The main remaining threats for this use are client authentication and non-repudiation. This paper presents a scheme to address SSL/TLS security holes, when it is used for web transaction security. The focus is only on transaction that is carried out by using credit/debit cards. The scheme uses wireless public key infrastructure (WPKI) in the client's mobile phone to generate a digital signature for the client. Thus we obtain client authentication and non-repudiation. At the same time, no overhead is imposed on the client, there is no need for any change to the actual system when performing the transaction, and no connection, by using the mobile phone, is required to perform the transaction.

Keywords: Web transaction security, SSL/TLS, digital signature, wireless security services.

1   Introduction

Security is a major concern for all involved in e-Commerce and particularly in the case of web transactions using debit/credit cards. Although the secure electronic transaction (SET) protocol [10] is able to achieve all the security requirements for web transactions, the implementation difficulties have made it redundant and now the preferred method is that of using the secure socket layer (SSL) [3] or transport layer security (TLS) [2]. In particular, SSL/TLS already exists in most web browsers and does not require any special effort from the client. The main security services which SSL/TLS can provide include [5,9,11] :

- Authentication of the merchant by means of his public key certificate.
- Data confidentiality during the transaction by encrypting the exchanged data.
- Data integrity for transactions enabled by using a digital signature.

From the security requirements for web transactions we observe that SSL/TLS does not fulfill all the security requirements, in particular:

G. Psaila and R. Wagner (Eds.): EC-Web 2007, LNCS 4655, pp. 11–20, 2007.

- Authentication for the client. In SSL/TLS, it is optional and typically, it is difficult for a normal client to have a public key certificate. As a result, any one who knows the credit card details is able to complete the transaction on behalf of the client.
- Non-repudiation. Neither the client nor the merchant has any proof of the transaction. The two parties are able to repudiate the transaction or change the transaction details after it has been carried out.
- SSL/TLS provides data confidentiality and data integrity only during the transaction. After completing the transaction, even with a legitimate merchant, all the client's details will be in the merchant's database. Any security breach of this database will lead to the disclosure of all the transaction details.

To eliminate the security risks associated with SSL/TLS, a SSL/TLS connection should be built on mutual authentication i.e. the merchant and client must have public key certificates. If every client has a public key certificate then full public key infrastructure (PKI) should be available and implemented. From previous experience (SET for example), any protocol which uses public key cryptography (PKC), that assumes the availability and full implementation of PKI, will not succeed. This paper presents a scheme to support SSL/TLS as a means for web transaction security. The scheme uses the mobile phone as a tamper resistant and calculation device to store the client's private key and to calculate a digital signature.

2   Using a Digital Signature for Web Transaction Security

In this section the proposed protocol is described. We will first start by the registration or setup phase and then describe the transaction phase. In the scheme the client must have a device to generate a digital signature that we call a digital signature creating device (DSCD), which will be discussed later in Section 3. At this stage we assume the client has a DSCD and she must use it to complete a transaction.

2.1   Registration Phase

This phase is part of the usual procedure which the client must follow to open a bank account and get a debit/credit card from the issuing bank (issuer). In this phase the client submits her credentials such as proof of her identity, address, etc. In addition to the personal details, the client submits her public key to be part of her credentials. The client can generate a pair of public/private keys for herself or through any trusted third party. The client must only use the public key parameters, which are acceptable by the issuer. It would be a good practice to let the issuer generate the cryptographic keys for the client. The private key must be generated and transferred to the client by an authentic method, such as is used today to generate the personal identification number (PIN) for debit/credit cards.

After the keys generation, the client stores the private key in her DSCD and submits her public key to the issuer and there is subsequently no need for a digital certificate for the public key. As part of the trust for the public key during the registration phase, a test for this key is carried out. This test could be carried out by signing a message using the private key in the DSCD and then checking the signature using the public key. In other words, the trust for the public key is part of the trust for the client, which is conducted by her credentials. The issuer must keep the client data confidential and maintain its integrity. The public key needs data integrity only while confidentiality is not required. The client is able to change her keys pair and submit a different copy of her public key at any time in the same way as changing any other personal detail. By finishing this phase the client is ready to use the protocol.

2.2   Transaction Processing

The client browses the merchant's web site and when she decides to pay for goods or services, a secure SSL/TLS connection, based on the merchant's public key certificate, is established. The protocol's steps can be summarized as follows. (see Figure 1).

1. The merchant sends a purchase form (OI) to the client. This purchase form must include the transaction value, date, merchant's identifier, and a unique transaction identifier. In addition to these fields, there is a field which we will call the *challenge* field. This field contains a hash function (HOI) for some selected values from the purchase form. There is also another empty field which we will call the *response* field.
2. The client authenticates herself to DSCD, enters the *challenge* value (HOI) into her DSCD and computes a digital signature (sig). The DSCD will generate a response for the *challenge*. The client fills in the signature value (sig) to the *response* field.
3. When the client submits the purchase form, she will receive the card details form from the merchant.
4. The client fills in her credit card details (PI) and submits the form to the merchant.
5. The merchant submits the selected data from the purchase form (OI), the *challenge* value (HOI), the *response* value (sig), and the debit/credit card form (PI) to the issuer via the acquirer. The issuer must be able to check the validity of the *challenge* value, i.e. the *challenge* value is really a hash function for the selected received values from the purchase form. At the same time, the merchant should not send the goods or service details to the issuer to maintain the privacy of the client.
6. The issuer checks the card details and uses the client's public key to verify the *response*. If these checks are correct the issuer authorizes the payment (Authorization).
7. The merchant informs the client with the result of the transaction (Confirmation) and the transaction is completed.

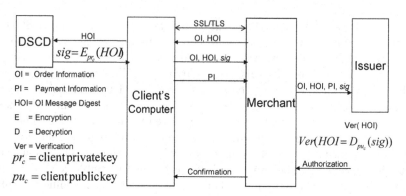

Fig. 1. Transaction process

2.3   Analysis

In this section we examine how much the new protocol covers the security holes of the SSL/TLS protocol, mentioned in Section 1.

- Authentication for the client. Since only the person who has the private key is able to generate the digital signature, so only the client, who has access to the DSCD, is able to complete a transaction. Thus authentication is achieved.
- Non-repudiation. The transaction details in addition to the merchant's identifier are joined together with their hash function (the *challenge* value) and signed by the client (the *response* value), so it is impossible to change any of these values after the transaction has been carried out. The client cannot repudiate the transaction because there is a copy from her digital signature joined up with the transaction. Hence, non repudiation is achieved.
- Data integrity and confidentiality after the transaction. Data integrity is provided after the transaction, as discussed above. The confidentiality of the client's data depends on the confidentiality of the merchant's database, which contains the transaction data. However, this is better than naive SSL/TLS because the user authentication value is not part of this database. If the protocol is widely used and the transaction is not possible without client authentication, then an attacker, who has access to this database, is not able to complete a transaction on behalf of the client.

As a result, by using a digital signature from the client in addition to SSL/TLS built on the merchant's public key certificate, all the security requirements for web transactions have been satisfied.

In practice, there is still a possibility of attack. If a fraudulent merchant changes some values (the price for example) in the OI so the merchant will have fake OI (FOI). Since the merchant generates the hash value for the transaction, the *challenge*, the merchant can generates a hash value for FOI (HFOI) and sends OI and HFOI to the client for signing. Because checking the hash function is too hard for the client, the client will check only the OI and sign the value HFOI. The merchant will submit FOI, HFOI, and the signature to the issuer. As we can observe, the issuer will find all these values are correct and will authorize FOI. To deter the merchant from this attack, the client must be able to take a soft or hard copy of the order form with the *challenge* and *response* on it. This is equivalent to taking a copy from a contract, in the real world.

3   Digital Signature Creating Device (DSCD)

The DSCD must be able to attain two main objectives. First of all, it needs a secure method to store the client's private key, and secondly, it needs enough processing power to generate the digital signature. Normally, the private key is stored in the client's computer, which could be used to generate the digital signature. But this method has many disadvantages such as destroying the mobility of the client i.e. the client must use the same computer to perform a transaction. There is also the possibility of the private key being stolen by unauthorized access to the computer or by malicious software. Another possibility is to use flash memory to save the keys. The access to the keys can be protected by a password. The method is good because it maintains the mobility of the client but the keys must leave the flash memory to the computer's memory and as a sequence it is still prone to attack by malicious software.

Hiltgen, et. al. [6] proposed the following factors to protect the private key. The private key should be stored in a tamper resistant hardware token such as a microprocessor-based smart card. The smart card should not be connected to the computer and the connection with the smart card should only be achieved via a smart card reader with a keyboard and display. We can observe that a scheme, which uses a smart card with a smart card reader, is secure but it has one major disadvantage which is the need for a smart card reader to perform the transaction. In the next sections we will present a new solution which is able to fulfill the conditions mentioned above and does not add too much overhead to the client.

4   Mobile Phone Security

Today, the mobile phone has become ubiquitous. In every mobile phone there is a subscriber identity module (SIM) which is essentially a smart card and the mobile phone acts as a smart card reader. The SIM contains the necessary information to authenticate the mobile phone to the mobile network. The research in this area, which sets up the mobile phone to be a tool for a secure connection, has a lot of promise because the mobile phone is used to surf the Web and it is necessary to have the ability for a secure connection.

The wireless application protocol (WAP) [15] is a protocol stack for wireless communication and is similar to the Internet protocol (IP) stack. Our interest in the WAP is focused on the security services. WAP security services include wireless transport layer security (WTLS) and application level security (ALS). WTLS [16] is a light version of TLS and is able to achieve the same security services. ALS includes some wireless PKI functions, the most important of which is the signText() function, defined in the Crypto library [17]. signText() generates digital signatures for the data.

Any parameter of mobile security such as a private key or a session key should be stored in tamper resistant hardware, called the wireless identity module (WIM). In addition to storing the security parameters, any use of these security parameters must be performed in WIM. WIM can be the SIM card itself or it can be another smart card. In the latter case, the mobile phone needs a second slot to install the WIM. Any access to the security functions in the WIM is protected normally by a PIN code. The client can change it at any time, and it is stored in the WIM and cannot be tampered with.

5   Mobile Phone as a DSCD

In this section we discuss how the mobile phone can be used to serve as a DSCD for the protocol mentioned in Section 2. After the client submits her details a *challenge* will be generated. The signText() function, which is used for the signature, is defined as follows [17]:

$signedString$ = Crypto.signText(*stringToSign, options, keyIdType, keyId*)

stringToSign: is the text which will be signed, the *challenge* in our protocol.

options: contains several options for the output such as if the *stringToSign* or a hash function from the public key, which will verify the signature, will be included in the output. In our protocol the options value should be zero.

keyIdType and *keyID*: these two fields are used to identify the public key, which will verify the signature. For example, *keyIdType* may indicate that *keyId* contains a hash function of the public key. In our protocol these two fields are zero. *signedSting* is a string type. It contains the output of the signing function. To summarize, we can write:

$signedString$ = Crypto.signText(*challenge,0,0,0*)

The *response* value consists of *signedString* in addition to some parameters such as the signing algorithm code, the certificate type, and the public key certificate. In our case, where no certificate will be sent, these parameters can be minimized to a few bytes. Thus, the *response* is nearly the same size as *signedString*.

As mentioned above, the *challenge* value is the output of a hash function. The most popular hash function nowadays is the secure hash algorithm 1 (SHA-1) [4]. The output of SHA-1 is 160 bits. The *response* is the result of a digital signature algorithm. The digital signature algorithm (DSA) and the elliptic curve DSA (ECDSA) require twice 2n bits for a security factor of n. Hence, if we require a security equivalent to a 64-bit key search, which is enough for our application, we need twice 128 bits of signature or 256 bits in total.

Crypto library uses Base-64 encoding to encode the *response*. In Base-64 encoding [1] every 6 bits need one character, thus the *response* will be a stream of 43 characters. If we use the same encoding for the *challenge* we need a stream of 27 characters. If the client has to copy one of these streams between the computer and her mobile phone, first it will be inconvenient, and second there is a high probability of error.

Today, mobile phones are not only used to make phone calls, but also to store and transfer data. Therefore the methods to connect the mobile phones to the computers are very popular. Examples of such methods are Bluetooth, cable, infrared, and memory cards. By using one of these methods, the client can send the *challenge* from her computer to the mobile phone, generate the digital signature and send the *response* to the computer to complete the transaction. However, if the client prefers to exchange the *challenge* and *response* values with the merchant by other means such as a short message service (SMS), then an extra option can be added and the protocol is still applicable.

Asymmetric cryptography needs a lot of memory and is processor intensive and both are not available in mobile phones. The main asymmetric operations for WAP security are WTLS connection, digital signature generation, and digital signature verification. A WTLS connection contains many generations and verifications for digital signatures, so it is much harder than a single digital signature generation or verification. Digital signature verification is harder than generation because to verify a signature signed data must be parsed and split into data and signature parts, the public key must be read from the certificate and the signer's certificate must be validated [7].

The protocol uses only one digital signature generation so there is no excessive use for asymmetric cryptography. Furthermore, the data which will be signed is already hashed by the merchant, the *challenge* value in the protocol. In some cases the hash algorithm is integrated with the sign algorithm, even in this case the extra hash will not add too much overhead because the data is so short. If the calculation is carried out by using WIM, the calculation can be accomplished. The research carried out by Weigold [14] showed that 200 milliseconds are required to calculate the digital signature.

In the market today many mobile phones are ready to perform digital signature generation, examples of which are: the Sony Ericsson (T610 and T68i) and the Nokia (6170 and 6800,). To sign text, by using one of these mobile phones [8], the user should use her mobile phone to browse the merchant web page. The client selects a link to the required goods. The text to sign will be shown and this text includes the goods specification, price, date, etc. The client enters her signature PIN and confirms the signature. The signature will be generated and sent to the merchant.

Unfortunately, neither of the mobile phones in the market today, to the best of the author's knowledge, would be possible to control the input and the output of the signText() function, i.e. we cannot enter a value to the mobile, perform a digital signature to this value, and get the result. The control over the text, which will be signed, is not only required for our protocol but also for any protocol that uses the mobile phone as a general purpose DSCD.

A discussion of the features, advantages, and disadvantages of using the protocol with discussion for other possible options is presented below.

- By using a mobile phone to generate the digital signature, a cover for SSL/TLS security holes is achieved, as mentioned in the discussion in Section 2.3. In particular, there are now two factors in authenticating the client i.e. the client must have the mobile phone in addition to knowing her PIN code. At the same time there is no overhead on the client because the client already has (and always carries) her mobile phone. The only inconvenience is in exchanging some data between the computer and the mobile phone.

- No change is required to the mobile phone or the SIM card. The only need is to implant the Crypto library, which is already standard as part of the WAP security and already implemented in many mobiles phones in the market.

- As we have discussed in Section 2, the authentication for the public key is achieved by means of the client's credentials to the issuer during the registration phase. This gives an advantage that there is no extra cost on the client to obtain a public key certificate from a certificate authority, at the same time there is a disadvantage that the client cannot use her mobile as a general means to sign electronic documents with any other party. Nevertheless, if the client wants to use her mobile phone as a general purpose DSCD, she is still able to have a certificate for her public key and the protocol is still valid.

- From the merchant's view, the response value does not provide any authentication of the client, thus the merchant should wait for the authorization from the issuer to supply the service to the client. However, this is not a big drawback because the merchant should anyway check with the issuer the validity of the client's payment information.

- The generation of the digital signature is carried out locally, by using the mobile phone itself. Consequently, there is no need for any extra communication costs as would be the case if a mobile phone were used to communicate with a third party to perform the authentication.

- There is no need to change the software or hardware in the actual method of transaction. The only requirement is to add two fields to the purchase form to enter the *challenge* and *response*.

- If the client's computer is infected by a virus or Trojan horse, the virus or Trojan horse can collect the transaction details, including the credit card details. However, they are not able to reveal any information about the client's private key. Actually, the *challenge* and *response* values are not confidential. Viruses or Trojan horses can only deny the transaction to be completed by changing the integrity of the *challenge* or *response* values because it will result in a bad signature during the verification. This result is inevitable and can be achieved by other means such as changing the credit card number.

- The client can use her private key for strong authentication in other applications, such as a direct connection to the issuer, as is the case with internet banking. It is also possible for the client to submit her public key to another server and use the protocol above for authentication to that server. No security threat is raised in using the private key to generate a digital signature for more than one application.

– It is possible to share a secret key. Instead of the client storing a private key in her WIM and submitting her public key to the issuer, she shares the same secret key with the issuer. Such a scheme is presented by VISA [13]. In fact, this method is used between the SIM and the mobile home network, which provides the SIM, to authenticate the mobile phone to the mobile network. Sharing a secret key has many advantages such as; firstly, encryption operations by using the secret key are easier and need less memory and computation power compared with PKC. Secondly, the encryption input and output (the *challenge* and *response*) can be reduced to a few bytes, thus there is no need for any connection between the mobile phone and the computer, as the client can enter them manually. At the same time, there are many disadvantages for using a shared secret. Firstly, the client cannot use the same key to authenticate herself to other systems. Although, it is possible to store more than one shared secret in WIM, it is difficult for the client to manage these keys. Secondly, and this is the crucial disadvantage, there is no standard to store and access a shared secret in WIM [13].

6 Related Works

There are many schemes that use the mobile phone for authentication, examples of these schemes with brief review.

3-D Secure by VISA [13]. VISA presented many schemes to use the mobile phone for authentication, such as using a password, a shared secret key, and a public key. The schemes are good and have many ideas for further research but there is a major disadvantage that there must be a direct connection between the client and the issuer during the transaction, so that the client can authenticate herself directly to the issuer. This connection results in extra costs in addition to no solution being presented in the case of where a connection is not available between the client's mobile phone and the issuer. This maybe the case if the client is in a different country during the transaction.

The electronic ID (eID) scheme by Valimo [12]. The scheme uses the mobile phone as a DSCD for general use, i.e. there is a public key certificate for the public key and this certificate is obtained from a government authority (the police). The client can use the digital signature in the same way as a conventional hand written signature. The scheme is good and implemented in Finland. The main disadvantage is that there must be a connection to the mobile service providers (Valimo) to perform an authentication.

7 Conclusion

The simplicity of SSL/TLS makes it the preferable protocol for online transactions security, although it is not able to achieve all the security requirements. The actual use of SSL/TLS, which does not require a client's certificate, raises many risks such as the client authentication and non-repudiation. Performing a digital signature, by using the client's private key, makes it possible to overcome these security threats. However, a digital signature needs a pair of private/public keys. The public key needs a certificate from a trusted third party, and the private key needs a secure method for storage. That makes it undesirable and difficult to adopt for most clients, because it will add too much overhead for the client.

By using the client's mobile phone as a means to generate the digital signature, a safe place to store the private key has been achieved. Furthermore, the mobility of the client is still maintained because the client always carries her mobile phone with her. The certificate for the public key from a trusted third party is not required because the public key in the protocol must be trusted only by the issuer. This trust is achieved between the issuer and the client during the registration phase, which is part of the procedure to open a bank account and obtain a debit/credit card. The protocol does not require any change to the mobile phone or any additional software or hardware. It only requires an implementation of the wireless security services, which is already standard and implemented successfully in many applications.

References

[1] Borenstein, N., Freed, N.: MIME (Multipurpose Internet Mail Extensions), Part One: Mechanisms for Specifying and Describing the Format of Internet Message Bodies. RFC 1521, IETF (1993)
[2] Dierks, T., Allen, C.: The TLS protocol. ver. 1.0. RFC 2246, IETF (1999)
[3] Freier, A.O., Karlton, P., Kocher, P.C.: The SSL protocol. ver. 3.0. Netscape (1996)
[4] Ferguson, N., Schneier, B.: Practical cryptography. Wiley, Indian (2003)
[5] Hassler, V.: Security Fundamentals for E-Commerce. Artech House, Massachusetts (2000)
[6] Hiltgen, A., Kramp, T., Weigold, T.: Secure Internet Banking Authentication. IEEE Security and Privacy 4(2), 21–29 (2006)
[7] Klingsheim, A.: JABWT and SATSA. NoWires Research Group, Department of Informatics, University of Bergen (2006)
[8] Nokia 6170 user guide,
http://nds1.nokia.com/phones/files/guides/Nokia_6170_UG_en.pdf
[9] O'Mahony, D., Peirce, M., Tewari, H.: Electronic payment system for E-Commerce, 2nd edn. Artech House Publishing, Massachusetts (2001)
[10] SETCo: Secure Electronic Transaction Standard- Book, pp. 1–3 (1997)
[11] Stallings, W.: Cryptography and network security principle and practice, 4th edn. Prentice Hall, New Jersey (2006)
[12] Valimo LTD: Mobile Signature services-improving eID, http://www.Valimo.com
[13] Visa International Service Association: 3-D Secure Mobile authentication scenario. ver. 1.0 (2002)
[14] Weigold, T.: Java-Based Wireless Identity Module. University of Westminster, London, UK; IBM Research Laboratory, Zürich, Switzerland (2002)
[15] Wireless Application Protocol Architecture Specification,
http://www.openmobilealliance.org/tech/affiliates/wap/wapindex.html
[16] Wireless Transport Layer Security Specification,
http://www.openmobilealliance.org/tech/affiliates/wap/wapindex.html
[17] WMLScript Crypto API Library: WAP-161-WMLScriptCrypto-20010620,
http://www.openmobilealliance.org/tech/affiliates/wap/wapindex.html

XℓPPX: A Lightweight Framework for Privacy Preserving P2P XML Databases in Very Large Publish-Subscribe Systems

Angela Bonifati[1] and Alfredo Cuzzocrea[2]

[1] ICAR Inst., Italian National Research Council, Italy
bonifati@icar.cnr.it
[2] DEIS Dept., University of Calabria, Italy
cuzzocrea@si.deis.unical.it

Abstract. The problem of supporting privacy preservation of XML databases within very large publish-subscribe systems is rapidly gaining interest for both academic and industrial research. It becomes even more challenging when XML data are managed and delivered according to the P2P paradigm, since malicious accesses and unpredictable attacks could take advantage from the totally-decentralized and untrusted nature of P2P networks. In this paper, we propose **XℓPPX**, a distributed framework for very large publish-subscribe systems which supports (*i*) privacy-preserving fragmentation of XML documents stored in P2P XML databases, and (*ii*) the creation of trusted groups of peers by means of "self-certifying" XPath links. Furthermore, we present algorithms for querying privacy-preserving XML fragments in both schema-aware and schema-less mode, which are common scenarios when P2P XML databases operate in very large publish-subscribe systems. Finally, we complete our analytical contributions with an experimental study showing the effectiveness of our proposed framework.

1 Introduction

Supporting the privacy preservation of XML databases is a novel and interesting research challenge. With similar attractiveness, very large publish-subscribe systems are rapidly gaining momentum as long as innovative knowledge processing and delivery paradigms like Web and Grid Services take place. In such scenarios, due to its well-understood features, XML is widely used as basic language for both representing and processing semi-structured data located in remote databases within distributed and heterogeneous settings. For the sake of simplicity, here and in the following, we model an XML database as a (large) collection of XML documents on top of which traditional DBMS-like indexing and query functionalities are implemented, thus adopting the so-called *native XML databases*.

It is widely-recognized that providing solutions to guarantee privacy preservation over sensitive XML data plays a critical role in next-generation distributed and *pervasive* applications, and, particularly, in the context of very large publish-subscribe systems. This issue is even more challenging when XML data are managed and delivered according to the popular P2P paradigm, since malicious accesses and unpredictable attacks could take advantage from the totally-decentralized and untrusted nature

G. Psaila and R. Wagner (Eds.): EC-Web 2007, LNCS 4655, pp. 21–34, 2007.

of P2P networks. This scenario gets much worse when these networks admit *mobile peers*, since mobile devices have limited power and resource capabilities thus they cannot process huge amounts of data, neither implement complex *security counter-measures*. On the contrary, in publish-subscribe systems, the information is often carried by *data providers* that tend to enclose it into very large XML databases.

Starting from these considerations, in this paper we address the issue of efficiently querying P2P XML databases in very large publish-subscribe systems while guaranteeing privacy preservation over sensitive XML data. The solutions we propose to this challenge are codified inside the core layer of a lightweight distributed framework, called **X*ℓ*PPX**, which effectively and efficiently supports the representation and management of privacy-preserving P2P XML databases in very large publish-subscribe systems. In more detail, **X*ℓ*PPX** supports (*i*) the secure fragmentation of XML documents in very large publish-subscribe systems by means of lightweight XPath-based identifiers, and (*ii*) trusted groups of peers by means of secure XPath links that exploit the benefits deriving from well-known *fingerprinting techniques* [22]. Due to the latter feature, we name as "self-certifying" our innovative XPath links. Also, **X*ℓ*PPX** embeds algorithms for querying secure XML fragments in both schema-aware and schema-less mode, which is very useful in P2P networks.

In the rest of the paper, for the sake of simplicity we refer to privacy-preserving XML documents in terms of "secure documents". However, privacy preservation is quite different from both *security*, which deals with cryptography-driven algorithmic solutions for information hiding, and *access control*, which concerns with the problem of limiting information access to particular classes of end-users according to a pre-fixed access scheme (e.g., like in multimedia databases). By contrast, privacy preservation is mainly related to the problem of guaranteeing the privacy of sensitive data during data management tasks (e.g., query evaluation). Note that this problem delineates a double-edged sword, since data providers would make available knowledge and intelligent tools for efficiently processing knowledge, but, at the same time, they would hide sensitive knowledge (e.g., this could be the case of *Business Intelligence* systems). Conversely, data consumers would acquire useful-for-their-goals knowledge, whereas malicious data consumers would also access sensitive knowledge (e.g., this could be the case of attackers stealing enterprise's secrets).

2   Application Scenario and Motivating Example

We assume a very large publish-subscribe system in which peers can freely roam and extract knowledge from distributed XML databases located at particular nodes in the system (see Figure 1). Precisely, the system consists of a set of data providers holding XML documents, and a set of *data replicators* maintaining *views* on top of those documents. XML data stored by providers are yielded according to popular publish-subscribe dynamics (e.g., corporate B2C systems). Moreover, data providers notify data replicators when new documents of interest become available. On the other hand, data replicators build their views on the basis of locality (e.g., frequent queries and downloads), and periodically refresh such views. Mobile peers of the network are under the scope of a given data replicator, thus letting define several *domains of*

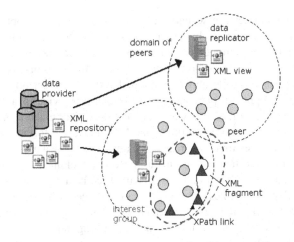

Fig. 1. Application scenario

peers. Domains of peers are freely built on the basis of (*i*) the interest of communities of peers in accessing views storing particular information (e.g., stock quotations), and (*ii*) locality issues (e.g., neighborhood criteria). Furthermore, without any loss of generality peers can also freely roam from one domain to another. It should be noted that the described setting covers a large set of modern systems and applications based on the novel Web and Grid Services paradigm, and, in particular, those adhering to the pioneering pervasive computing initiative.

Peers of such system are usually not interested to the entire view published by the local replicator, but to point some data contained into it, i.e. extracting *fragments* of the view. To avoid excessive computational overheads, peers prefer to access data replicators rather than data providers, while still wishing to access data providers whenever needed (e.g., due to load balancing issues). While peers fetch their fragments of interest, they also keep links to the related fragments stored on their neighbors. A link is represented as an absolute XPath expression within the original document, being such expression capable of uniquely determining the related fragment via prefix-matching. Peers linked to each other form an *interest group* (also called *acquaintance* in [3]), which can be viewed as a group of peers sharing *semantically-related fragments*, i.e. fragments related to concepts having a certain relationship with respect to the semantics of a specific application domain. Without any loss of generality, the same peer may belong to multiple interest groups.

In such a scenario, a pertinent problem is how to guarantee privacy preservation while accessing sensitive fragments, i.e. to avoid that untrusted peers can access secure fragments. It should be noted that while privacy preservation issues concerning a given data provider/replicator can be handled in a centralized manner by adopting specialized solutions, devising privacy-conscious P2P XML data processing solutions in environments like the one described above is still an open and unsolved problem, due the same nature of P2P networks. As a consequence, in this paper we focus on the latter research challenge.

In our reference application scenario, a peer has basically two options when it is looking for further information. In case it needs update its information, it has to access again the replicator, although consuming bandwidth and resources. In turn, when it needs *static data*, i.e. data with low variability over time, it is also offered to access its neighbor fragments via the above-mentioned links. Consider an example of this scenario in which a replicator publishes a view on public services available in a given urban area. Among the others, this view contains information about the local market stocks and the public transportation timetable. Notice that the former data are subject to a great variability over time, whereas the latter remain unchanged for a long time (e.g., a season). Imagine that a (mobile) user Bob is seeking for information about both kinds of data. For stock trends, Bob will directly access the local replicator, whereas for the departure times of trains, he can still access (trusted) neighbor's data (e.g., Alan, who had previously downloaded the same fragment of interest). In our framework, such a strategy is feasible thanks to our fragmentation model, which lets build links between related fragments.

A trusted peer can access the others' secure fragments by both directly browsing the links, and, alternatively, formulating an arbitrary XQuery query on (*schema-aware*) documents (note that fragments of documents are, in turn, documents). A further capability of our framework lets exploit the set of path expressions of an interest group to aid query formulation in the absence of a schema (i.e., against *schema-less* documents). It should be noted that schema-less query processing is a common situation in highly dynamic P2P networks, since, as shown in [3], a *global mediated schema* is not a reasonable assumption for such networks. For instance, Alan can retrieve the sets of path expressions of his interest group and use them in two ways: *(i)* grasp the fragments owned by other peers that may contain data he is interested to query later on, or *(ii)* use these path expressions to formulate his query, whenever the schema of the global document is not available.

As a side remark, note that handling updates in our framework (e.g., linked fragments that change in their structure or content), is outside the scope of this paper, and it is also an interesting and exciting research challenge we plan to address in the near future.

3 The XML Fragmentation Model in a Nutshell

For what concerns the P2P data layer, our framework relies on an *XML fragmentation model for P2P networks*, which is based on our preliminary work [4]. There, we assume to deal with P2P XML databases storing fragments of XML documents. In the present paper, we study the specific issues of hiding fragments to untrusted peers, thus protecting possibly sensitive data. We adopt a distributed, privacy-preserving mechanism according to which each (self-certifying) XPath link that points to a fragment is *(i)* encrypted by using a trusted key founding on the fingerprinting technique, and *(ii)* shared by all the other members of the same (interest) group. This solution also makes perfect sense when compared with *symbolic links* of *Self-certifying File Systems* (SFS) [12], since such links are encrypted by using a *Public Key Server* (PKS). In fact, in a P2P environment like the one we have just described in our application scenario (see Figure 1), we cannot rely on a central authority for public keys, due to fault tolerance and scalability reasons. Nevertheless, it is reasonable to make groups of peers sharing the *responsibility* of a key (details are given in Section 4).

Although popular P2P networks allow the users share entire data files, the P2P paradigm is flexible enough to be applied to data at any granularity. In particular, in [4] we study the issue of sharing XML fragments scattered to multiple peers. In this Section, we revise the concepts presented in that paper, which serves as background to the novel privacy issues presented here. The key concept behind our model is the fact that, in a P2P setting, an XML document *does not entirely* reside on one peer for either space or relevance reasons. Therefore, we focus on how to represent the fragments of a document split on several peers, in order to be able to re-unify those at wish. Hence, a fragment is a sub-document of the original document maintaining XPath links to the latter, thus retaining the convenient side-effect of building a *decentralized catalog* over the P2P network. In our model, we do not expect the peers to agree on one schema and keep it updated with respect to every neighbor's changes to local data. On the contrary, every peer has one or more fragments and it may execute global queries without necessarily knowing the global schema. Thus, as we said above, we can handle both schema-less documents, which are very common in P2P networks as well as the Web, and schema-aware documents, more common in distributed database systems. Our query mechanism only relies on links for the first kind of documents, while it also looks at the local version of the schema for the second kind. Notice that this local version may disagree with other versions present in the network, since, as we said previously, we are not assuming a common mediated schema among peers.

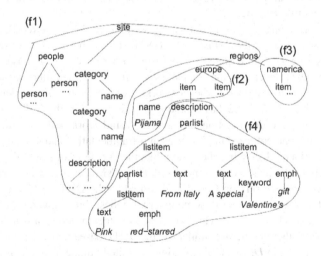

Fig. 2. A snippet of an XMark document

To enable fragmentation of XML documents, our model exploits a set of *lightweight path expressions*. A fragment f is a valid sub-tree of the original document having the following set of paths: (*i*) the *fragment identifier* f_i, i.e. the unique path expression identifying the root of the current fragment within the global document; (*ii*) the *super-fragment path expression* p_s, i.e. the unique path expression linking the parent of the current fragment; (*iii*) one or more *child fragment path expressions* p_c, i.e. the path expressions linking the children of the current fragment. While f_I and p_s are stored separately from the fragment content, p_c paths are stored within special tags sub added as fragment leaves. Schema-aware documents also store on each peer the XML schema of the documents along with the local fragments and the above-described paths. Figure 2 shows a snippet of an XMark document [23], shredded according to our fragmentation strategy. Fragment f_2 = /site/regions/europe has /site as super-fragment and /site/regions/europe/item[1]/description as child fragment. Similarly, fragment f_I = /site has no super-fragment and has both /site/regions/ europe and /site/regions/namerica as child fragments. Henceforth, paths sharing the same prefix, i.e. belonging to the same document, are named *related paths* in our fragmentation model.

Path expressions described above play a crucial role in identifying the fragments because they represent flat keys over the network. Moreover, they are more lightweight and scalable than B^+-trees used in [9] to answer range queries over relational data. In the following, we will describe how the fragments are allocated in a P2P network implementing **X***l***PPX**, thus adhering to our XML fragmentation model. Here, we want to highlight that our strategy is reminiscent of vertical partitioning in distributed database systems, but customized for XML data. This distribution strategy uses a limited fragment of XPath, namely *light XPath*, consisting only of / and [i], such that [i] indicates positional axes, which are useful to identify fragments. Note that the original document order is preserved in both assigning fragment identifiers and in encoding child fragments.

For the network configuration, in our setting we assume the structured P2P paradigm [13,15], which uses a *Distributed Hash Table* (DHT) to bias the search towards particular peers. Structured networks are quite different from super-peer networks used in [21]. In our model, the DHT leverages the uniqueness of fragment identifiers to identify and (al)locate the needed data over the network. For instance, a fragment such as /site/regions/namerica has a unique encoding within the DHT employed by the network. This path identification mechanism would work with any fragmentation technique (e.g., the one proposed in [5]), or even if fragmentation had not happened, such as, for instance, in a publish-subscribe system like the one described in [14] where entering machines publish their data by means of (Web/XML-enabled) path expressions. Implementation-wise, we rely on Chord [25] to realize the DHT. In Chord, each node keeps a logarithmic routing table to some external nodes, thus allowing a considerable network efficiency to be achieved. This is opposite to other systems, in which each node is connected to all other nodes, and similar to other DHT-based networks, such as *Pastry* [8] and *P-Grid* [1]. As in *DHash* [10], we admit a limited number of replicas of the fragments. Besides being stored on the current peer, the same fragment is replicated on every node of the peer routing table. This is done in order to guarantee the reliability of the network in the presence of peer failures.

Each fragment comes with the p_c and p_s path expressions, and with its fragment identifier f_I stored within the local peer. The fragment identifier is exposed to the outside under the form of a unique key. In such a way, any other peer that looks for a particular path expression will search it through the DHT. We have extended the Chord DHT to support this behavior. However, in the original Chord, the hash function used is SHA-1, which is replaced in our model with the fingerprinting technique. Fingerprinting path expressions in a P2P network is similar to fingerprinting URLs [6], but different from an application point of view. In [4], we provide an experimental evaluation demonstrating that hashing and fingerprinting guarantee the same load balancing. We prefer fingerprints to any arbitrary hash function because of their software efficiency, their well-understood probability of collision (discussed in [4]) and their nice algebraic properties.

Now, we discuss how fingerprinting works. Let $A = \langle a_1 a_2 \ldots a_m \rangle$ be a binary string. We associate to A a polynomial $A(t)$ of degree $m - 1$ with coefficients in the algebraic field Z_2, $A(t) = a_1 \cdot t^{m-1} + a_2 \cdot t^{m-2} + \ldots + a_m$. Let $P(t)$ be an irreducible polynomial of degree k, over Z_2. Given $P(t)$, the fingerprint of A is the following: $f(A) = A(t) \bmod P(t)$. The irreducible polynomial can be easily found following the method in [22]. Therefore, in order for a peer to compute the path expression fingerprint, it suffices to store the irreducible polynomial $P(t)$. The latter has a fixed degree equal to $NF + 2 \cdot DM + Q$, being (i) 2^{NF} the number of fragments in the network, (ii) 2^{DM} the length of the longest path expression in the network, and (iii) 2^Q a threshold due to the probability of collision between two arbitrary distinct tokens [6]. It should be noted that polynomials are quite small structures to be replicated on each participating peer if compared to *replicated global indexes* used in [5]. Moreover, our set of lightweight path expressions and the accompanying polynomial are not directly comparable to probabilistic approaches based on *bloom filters* as in [14], which can handle XML data of relatively small depth.

4   XℓPPX: Overview and Privacy Preservation Features

In order to support the privacy preservation of XML fragments over P2P networks in very large publish-subscribe systems (and, as a consequence, that of database storing such fragments), in **XℓPPX,** XPath links pointing to fragments are encrypted using a trusted key encoded by means of a fingerprinting technique, and shared by peers of the same (interest) group. As we said above, such links are thus identified as "self-certifying" links, i.e. able to guarantee privacy preservation features.

To this end, besides fingerprinting path expressions, we also *fingerprint the actual XML content of the fragments*. This is novel with respect to previous initiatives, and not discussed at all in [6], where fingerprinting is not used for authentication purposes. Indeed, since fingerprinting, like hashing, reduces any arbitrary string to a fixed-length token, we can safely apply fingerprinting to the serialized content of an XML fragment. Since all we need to decode a fingerprinted item is the irreducible polynomial $P(t)$ (see Section 3), it is straightforward to create interest groups that share the same polynomial. Every peer within such groups can verify the authenticity of fragments in the community, and contribute to any issued query, which would be blind to the others. Thus, we enable privacy-preserving query processing with such verified peers. Of course, there will be as many groups of peers as the number of polynomials we wish to allow in the network, ranging from the scenario with one distinct polynomial per-peer or per-group-of-peers to the scenario with one unique polynomial for all peers. Notice that this approach also guarantees that peers that answer queries are trustworthy (we describe our query algorithms in Section 5).

Figure 3 shows an overview of **X𝓁PPX**, where four peers and two interest groups are depicted. In more detail, the peers P_0 and P_3 form an interest group via sharing the polynomial $Poly_X$, whereas the peers P_1 and P_2 form another interest group via sharing the polynomial $Poly_Y$. The Figure also shows the logical architecture of the components that need to be implemented on each peer within **X𝓁PPX**. Consider a peer P_i among those depicted in Figure 3. Components implemented on P_i are the following: (*i*) *P2P Middleware*: it is the basic P2P middleware implementing traditional P2P communication protocols among peers of the network; (*ii*) *Privacy-Preserving Module*: it is the component of our framework implementing our fingerprint-based privacy-preserving scheme; (*iii*) *Fragment Manager*: it deals with the management of fragments according to our P2P XML fragmentation model; (*iv*) *Data Access Module*: it provides data access functionalities on the underlying P2P XML fragment database; (*v*) *Data Query Module*: it provides data query functionalities on the underlying P2P XML fragment database; (*vi*) *External Data Access API*: a collection of APIs used to access and query the local data replicator when up-to-date information is required (see Section 2).

Our proposal resembles the use of cryptography [17] (e.g., algorithms RSA and AES), and *self-certifying path names* [12], which are used in SFS directories to encrypt the host symbolic names. However, these cryptographic functions yielding large pieces of data introduce excessive spatio-temporal overheads in an overlay P2P network. Therefore, as claimed in Section 1, even if security issues obviously influence our work, our effort is in the privacy preservation of P2P data management.

Along with the basic idea of fingerprinting paths and fragments to support privacy preservation features, we also provide two extensions useful to handle privacy preservation among peers, which overcome more traditional schemes, and can be used to develop more specific privacy-preserving protocols for large and highly dynamic P2P networks. First, in traditional schemes (e.g., SFS), public key is decided by a centralized server, which would be not applicable in a P2P network. Contrarily to this approach, in order to decide the public key, we actually use an *authority group* policy as in [19]. More precisely, we require that *a public key is jointly decided by all the group members*. Secondly, traditional schemes focus on secure handling host names, whereas in our solution we need a broader level of security, i.e. extended to any (XPath) links either embedded in a fragment or lying out of it.

Handling new peers joining the network and wishing to be admitted in an interest group is another interesting issue, which becomes relevant with respect to the problem of avoiding that malicious peers attack the network. According to the authority group policy implemented in **X𝓁PPX**, the admission procedure requires that a new peer entering the network can be admitted in a group *if and only if* all the members agree on its admission. In more detail, in our model, this only applies if the new peer holds fragments related to those stored by other members. If this is not the case, then the peer can only be assigned to a new public key and create its own group.

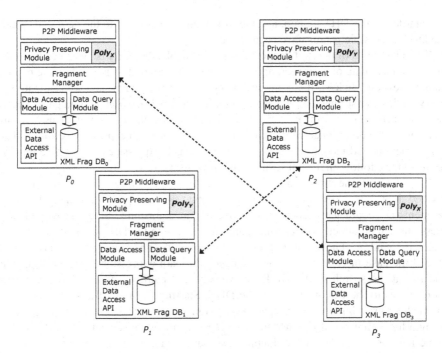

Fig. 3. XℓPPX Overview

There may be other kinds of attacks on a DHT-based P2P network, which go beyond the authenticity of data, such as threatens to the liveness of the system, and prevention of participants to find data (see [24] for a complete survey). These issues are orthogonal to the problem we investigate in this paper.

5 Schema-Less and Schema-Aware Evaluation of XPath Queries in XℓPPX

An arbitrary user can pose a query against the view of a replicator or, alternatively, against the fragments of the neighbor peers. We focus on the latter kinds of queries, as querying the replicator is quite straightforward and based on well-known solutions (e.g., query answering using views). To query linked fragments, our fragmentation model provides an effective strategy that inspects the information encoded into self-certifying path expressions. The latter represent a distributed catalog that can be fruitfully exploited during query evaluation. We base our description on examples that refer fragments shown in Figure 2. We discuss the evaluation of *descendant XPath queries*, since more complex XQuery queries use this module as a building block. While we use *light XPath* for fragment identifiers, the XML language we use for queries is complete XPath 1.0. Given an arbitrary XPath query, we divide it into several segments, either containing only / or containing // only. Note that, among all filters, only positional ones are kept in the path expression, as the others were not admitted as fragments identifiers. Thus, filters other than positional are initially disregarded in the DHT lookup, and their evaluation postponed to the very-end, when the DHT-accessible results have been found.

When a schema of the split document is not available, the query is evaluated by looking at the local data of each peer, and by guessing the external data based on the related path expressions. Indeed, if we are evaluating `//description` and we originate the query on the peer storing the fragment f_2, then we already learn from f_2 child path expression that there is a tag `description` on the peer storing the fragment f_4. This gives us a priority in the order links are followed: this "promising" path expression is followed as first. When instead the tags contained in the child path expressions do not give this information, we follow all the links without giving priority to any.

Consider now the query `/site/regions//item//description`, which is divided into `/site/regions` and `//item//description`. Hence, the two segments are evaluated according to two different algorithms. Single-slash path expressions can be directly fingerprinted and matched with the DHT content. In case no match is found, we know that no fragment is rooted in `/site/regions` (and this is indeed the case in Figure 2). Thus, we prune the path expression one step at a time and repeat the search in the DHT. Conversely, the path expression `/site`, which corresponds to the fragment f_1 is in the DHT. Starting from this fragment, our algorithm attempts to build the entire path expression `/site/regions`, with `regions` being a local element of f_1. Once the `regions` elements have been found, these become the context nodes of the double-slash path `//item//description`. Double-slash paths are evaluated one step at a time. To improve performances, the p_c paths of f_1 are followed in parallel to search item elements. The reader should note that here p_c paths do not let guess the presence of items on external fragments. The element `item` under `namerica`, as well as the rhs `item` under `europe`, is discarded as it has not neither local child nor external p_c paths to explore. Conversely, the lhs `item` under `europe` shows an external path which is promising in that it has `description` as final step. The latter path is then followed, and the element `description` is returned as result[1].

When a schema of the document is available, simple optimizations can be enabled. We emphasize here the use of the schema as a sort of *data summary*, similar to what provided by *data guides* and indexes for XML data [18]. In fact, the evaluation of both single-slash and double-slash queries can be improved by quickly determining and discarding ill-formed queries, i.e. queries that do not respect the available schema. For instance, a query like `/site//teacher` can be promptly declared as empty. The schema may also help in evaluation of queries containing wildcards, such as `/site//*/person`. In such a case, we avoid inspecting fragments f_2 and f_3 (and, as a consequence, f_4) if we can infer from the schema that `person` elements cannot lie underneath `regions` elements. The reader should notice that the two optimizations described above can be achieved by simply handling schemas of documents, thus without introducing excessive computational overheads. In any case, as highlighted above, schemas are not mandatory in our query strategy. Also, we expect that a peer modifies at will its local data by introducing new elements. We assume the modifications would not be communicated to other peers.

---

[1] We assume here the network only has these fragments, otherwise the search would exhaustively explore the other fragments till coverage of all the nodes.

6   Experimental Study

In order to test the effectiveness of our proposed framework, we conducted various experiments by using as input the popular XML benchmark data set XMark. As highlighted previously, we focus our attention on testing the capabilities of the P2P layer of **X𝓁PPX**, having the latter a more critical role than data-provider and replicator layers. In particular, we performed two sets of experiments. In the first one, we stressed the *query capabilities* of **X𝓁PPX**, i.e. the capability of **X𝓁PPX** in answering XPath queries over fragmented XML data, whereas in the second one we stressed the *privacy preservation capabilities* of **X𝓁PPX**, i.e. the capability of **X𝓁PPX** in supporting privacy preservation functionalities over fragmented XML data.

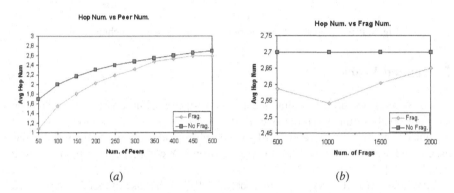

Fig. 4. Query performance on the XMark benchmark Xm_{30}

In our experimental setting, we use polynomials with degree equal to 64, which leads to an acceptable probability of 2^{-10}, and allows us to exploit a maximum length for path expressions of 50 steps (averaged on a length of 10 symbols per step), and a maximum number of fragments equal to 2^{30}, which is huge enough for arbitrary networks. Figure 4 and Figure 5 show the experimental results of our analysis, which has been conducted over an XMark document that was 30 MB is size, denoted by Xm_{30}.

For what concerns the query capabilities of **X𝓁PPX**, Figure 4 (*a*) shows the average hop number with respect to the number of peers in the two different settings in which (*i*) Xm_{30} is split into 1000 fragments, and (*ii*) Xm_{30} is kept entire. Figure 4 (*b*) shows the same metrics with respect to the number of fragments (the number of peers is set to 500). For what concerns the privacy preservation capabilities of **X𝓁PPX**, Figure 5 (*a*) shows the average collision number in percentage among fingerprints with respect to the number of fragments, when varying the number of bits used to encode fingerprints. Finally, Figure 5 (*b*) shows the same metrics with respect to the number of bits used to represent fingerprints, when varying the number of fragments in the network. In both latter cases, the number of peers is set to 500, and the number of interest groups is set to 5. From the analysis of the experimental results, it clearly follows that the proposed framework allows us to efficiently query P2P XML databases while guaranteeing privacy preservation over sensitive XML documents.

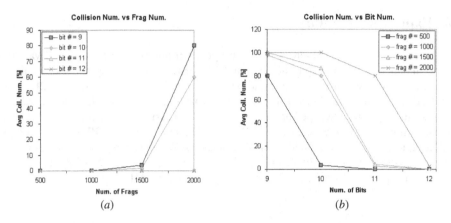

Fig. 5. Privacy preservation performance on the XMark benchmark Xm_{30}

7 Related Work

In this Section, we briefly outline some state-of-the-art solutions for supporting privacy preservation functionalities over P2P networks, which are close to our work. On the other hand, the literature on XML-based publish-subscribe systems is well established, and we omit it here, due to space reasons. A comprehensive tutorial can be found in [11].

Threshold distributed cryptography is used in [19] for the sake of membership control. [19] casts doubt on the viability of these approaches in unstructured P2P networks if they do not guarantee verifiability of group members compromised by adversaries. We acknowledge this issue, and focus on a simple yet effective solution for trusted P2P XML fragments. Security issues in an agent-based P2P network are examined in *BestPeer* [21]. However, security has a different meaning there, i.e. to protect the data carried by agents when they inspect fixed or unknown network paths. For this purpose, *signencryption* [17] is used. However, no experimental evidence is given to justify the scalability of cryptography in P2P networks. Our approach is different as it is customized for XML fragments over P2P networks. 128-bit encryption to protect peer communication is also used in agent-based *PeerDB* [20]. *PeerDB* differs from our framework in that it performs IR-based queries on non-distributed relational data. However, like ours, it realizes a P2P full-fledged database. Access-control on each peer is used in [2] to protect *active XML documents* containing service calls. Goals of [2] are different from ours, as [2] aims at providing a sort of "application-level" degree of security which could indeed used on top of our framework to exploit the available functionalities in the context of (secure) P2P Web Services. Security on semantic views of Web information are realized by means of *information mediators* in [16]. This architecture, although very interesting for a Web context, is not applicable to a P2P network. Handling uncooperative or selfish nodes in a large-scale P2P network is done in [7]. Both problems are very interesting but orthogonal to ours.

8 Conclusions and Future Work

In this paper, we have presented **XℓPPX**, a framework for supporting privacy preservation of XML fragments among peers embedded in very large publish-subscribe systems, where XML views are published by both data providers and replicators. Efficiently supporting query processing on secure P2P XML fragments is another goal of **XℓPPX**. In **XℓPPX**, privacy preservation features across peers are provided by means of well-established fingerprinting techniques, whose reliability and efficiency have been proved via a comprehensive experimental study over XMark documents. Schema-less and schema-aware query algorithms for the evaluation of descendant XPath queries over secure XML fragments have also been presented and discussed. The experimental evaluation of **XℓPPX** has further confirmed the advantages deriving from our fragmentation technique, and the effectiveness of the proposed privacy-preserving scheme. Future work is mainly focused on extending the query functionalities implemented in **XℓPPX**, in order to include more advanced IR capabilities. Moreover, we plan to test the scalability of privacy preservation features of **XℓPPX** in real-life systems.

References

[1] Aberer, K., Cudre-Mauroux, P., Datta, A., Despotovic, Z., Hauswirth, M., Punceva, M., Schmidt, R.: P-Grid: A Self-Organizing Structured P2P System. SIGMOD Record 32(3), 29–33 (2003)

[2] Abiteboul, S., Alexe, B., Benjelloun, O., Cautis, B., Fundulaki, I., Milo, T., Sahuguet, A., An Electronic Patient Record "on Steroids": Distributed, Peer-to-Peer, Secure and Privacy-conscious. In: Proc. of VLDB, pp. 1273–1276 (2004)

[3] Bernstein, P.A., Giunchiglia, F., Kementsietsidis, A., Mylopoulos, J., Serafini, L., Zaihrayeu, I.: Data Management for Peer-to-Peer Computing: A Vision. In: Proc. of ACM WebDB, pp. 89–94 (2002)

[4] Bonifati, A., Cuzzocrea, A.: Storing and Retrieving XPath Fragments in Structured P2P Networks. Data & Knowledge Engineering 59(2), 247–269 (2006)

[5] Bremer, J.-M., Gertz, M.: On Distributing XML Repositories. In: Proc. of ACM WebDB, pp. 73–78 (2003)

[6] Broder, A.Z.: Some Applications of Rabin's Fingerprinting Method. Springer, Heidelberg (1993)

[7] Buchmann, E., Bohm, K.: FairNet - How to Counter Free Riding in Peer-to-Peer Data Structures. In: Proc. of CoopIS/DOA/ODBASE, vol. 1, pp. 337–354 (2004)

[8] Castro, M., Druschel, P., Hu, Y.C., Rowstron, A.: Proximity Neighbor Selection in Tree-based Structured Peer-to-Peer Overlays. Technical Report MSR-TR-2003-52, Microsoft Research (2003)

[9] Crainiceanu, A., Linga, P., Gehrke, J., Shanmugasundaram, J.: Querying Peer-to-Peer Networks Using P-Trees. In: Proc. of ACM WebDB, pp. 25–30 (2004)

[10] Dabek, F., Brunskill, E., Frans Kaashoek, M., Karger, D., Morris, R., Stoica, I., Balakrishnan H.: Building Peer-to-Peer Systems with Chord, a Distributed Lookup Service. In: Proc. of IEEE HotOS, pp. 81–86 (2001)

[11] Jacobsen, H.-A., Llirbat, F.: Publish-Subscribe Systems. Tutorial at IEEE ICDE (2001)

[12] Fu, K., Kaashoek, M.F., Mazieres, D.: Fast and Secure Distributed Read-Only File System. Computer Systems 20(1), 1–24 (2002)

[13] Galanis, L., Wang, Y., Jeffery, S.R., DeWitt, D.J.: Locating Data Sources in Large Distributed Systems. In: Proc. of VLDB, pp. 874–885 (2003)

[14] Gong, X., Yan, Y., Qian, W., Zhou, A.: Bloom Filter-based XML Packets Filtering for Millions of Path Queries. In: Proc. of IEEE ICDE, pp. 890–901 (2005)

[15] Huebsch, R., Hellerstein, J.M., Lanham, N., Loo, B.T., Shenker, S., Stoica, I., Querying the Internet with PIER. In: Proc. of VLDB, pp. 321–332 (2003)

[16] King, R.: Security Maintenance Mediation: a Technology for Preventing Unintended Security Breaches. Concurrency and Computation: Practice and Experience 16(1), 49–60 (2004)

[17] Menezes, A.J., van Oorschot, P.C., Vanstone, S.A.: Handbook of Applied Cryptography. CRC Press, Boca Raton (1996)

[18] Milo, T., Suciu, D.: Index Structures for Path Expressions. In: Beeri, C., Bruneman, P. (eds.) ICDT 1999. LNCS, vol. 1540, pp. 277–295. Springer, Heidelberg (1998)

[19] Narasimha, M., Tsudik, G., Yi, J.H.: On the Utility of Distributed Cryptography in P2P and MANETs: The Case of Membership Control. In: Proc. of IEEE ICNP, pp. 336–345 (2003)

[20] Ooi, B.C., Tan, K.L., Zhou, A.Y., Goh, C.H., Li, Y.G., Liau, C.Y., Ling, B., Ng, W.S., Shu, Y.F., Wang, X.Y., Zhang, M.: PeerDB: Peering into Personal Databases. In: Proc. of ACM SIGMOD, p. 659 (2003)

[21] Pang, X., Catania, B., Tan, K.: Securing Your Data in P2P Systems. In: Proc. of IEEE DASFAA, pp. 55–61 (2003)

[22] Rabin, M.O.: Fingerprinting by Random Polynomials. Technical Report CRCT TR-15-81, Harvard University (1981)

[23] Schmidt, A., Waas, F., Kersten, M., Carey, M., Manolescu, I., Busse, R.: XMark: A Benchmark for XML Data Management. In: Proc. of VLDB, pp. 974–985 (2002)

[24] Sit, E., Morris, R.: Security Considerations for Peer-to-Peer Distributed Hash Tables. In: Druschel, P., Kaashoek, M.F., Rowstron, A. (eds.) IPTPS 2002. LNCS, vol. 2429, pp. 261–269. Springer, Heidelberg (2002)

[25] Stoica, I., Morris, R., Karger, D., Frans Kaashoek, M., Balakrishnan, H.: Chord: A Scalable Peer-to-Peer Lookup Service for Internet Applications. In: Proc. of ACM SIGCOMM, pp. 149–160 (2001)

Usability Analysis Framework Based on Behavioral Segmentation

Peter Géczy, Noriaki Izumi, Shotaro Akaho, and Kôiti Hasida

National Institute of Advanced Industrial Science and Technology (AIST)
Tsukuba and Tokyo, Japan

Abstract. Usability analysis tools are indispensable for improving efficiency of electronic environments including web based portals and services. The article presents a novel behaviorally centered usability analysis framework. The framework has been applied to elucidation of usability characteristic of a large corporate intranet with an extensive knowledge worker user base. Knowledge workers had a significant tendency to form elemental and complex browsing patterns, utilized relatively narrow spectrum of available resources, and generally exhibited diminutive exploratory behavior.

1 Introduction

"Nobody has really looked at productivity in white collar work in a scientific way." (Peter Drucker) [1]. Insufficient scientific evidence concerning knowledge worker productivity and their relevant metrics have been the focus of the recent managerial discourse [2]. Extensive deployment of IT infrastructure in corporate sphere brings novel opportunities as well as challenges. Although usability analysis of electronic environments have been attracting significant attention from e-commerce perspective [3], there is substantially inadequate body of knowledge concerning the usability of corporate intranet portals.

Approaches to usability analysis of web based environments range from behavioral observer studies to mining clickstream data. The observer based behavioral studies [4] are being replaced by methods allowing higher level of automation [5]. Research activity in usability studies have led to a wide range of models and metrics, however, only limited attempts have been made toward unification of diverging streams [6]. Novel trends in usability analysis favor methodologies permitting on-the-fly massive data and user processing. Large data volumes and timely processing impose stringent computational efficiency requirements on the usability analysis techniques.

The presented usability analysis framework effectively captures navigational, temporal, and behavioral dimensions of human interactions in electronic environments. It is efficiently implementable and expandable. Application of the framework to usability analysis of a large corporate intranet with substantial knowledge worker user domain revealed several important attributes.

G. Psaila and R. Wagner (Eds.): EC-Web 2007, LNCS 4655, pp. 35–45, 2007.
© Springer-Verlag Berlin Heidelberg 2007

2 Approach Formulation

We introduce the basic line of inquiry together with the corresponding terminology. Definitions are accompanied by intuitive explanations that help us better understand the concept at a higher formal level.

The clickstream sequences [7] of user page transitons are divided into sessions, and sessions are further divided into subsequences. Division of sequences into subparts is done with respect to the user activity and inactivity. Consider the conventional time-stamp clickstream sequence of the following form: $\{(p_i, t_i)\}_i$, where p_i denotes the visited page URL_i at the time t_i. For the purpose of analysis this sequence is converted into the form: $\{(p_i, d_i)\}_i$ where d_i denotes a delay between the consecutive views $p_i \to p_{i+1}$. User browsing activity $\{(p_i, d_i)\}_i$ is divided into subelements according to the periods of inactivity d_i.

Definition 1. *(Browsing Session and Subsequence)*
Browsing session is a sequence $B = \{(p_i, d_i)\}_i$ where each $d_i \leq T_B$. Browsing session is often in the further text referred to as simply a **session**.
Subsequence of a browsing session B is a sequence $S = \{(p_i, dp_i)\}_i$ where each delay $dp_i \leq T_S$, and $\{(p_i, dp_i)\}_i \subset B$.

The sessions delineate tasks of various complexities users undertake in electronic environments. The subsequences correspond to the session subgoals; e.g. subsequence S_1 is a login, S_2 – document download, S_3 – search for internal resource, etc.

Important issue is determining the appropriate values of T_B and T_S that segment the user activity into the sessions and subsequences. The former research [8] indicated that the student browsing sessions last on average 25.5 minutes. However, we adopt the average maximum attention span of 1 hour for T_B. Value of T_S is determined dynamically and computed as an average delay in a browsing session bounded from below by 30 seconds: $T_S = max\left(30, \frac{1}{N}\sum_{i=1}^{N} d_i\right)$. This is preferable in environments with frame-based and/or script generated pages where numerous logs are recorded in a rapid transition.

Another important aspect is to observe where the user actions are initiated and terminated. That is, to identify the starting and ending points of the subsequences, as well as single user actions.

Definition 2. *(Starter, Attractor, Singleton)*
Starter is the first navigation point of an element of subsequence or session with length greater that 1. **Attractor** is the last navigation point of an element of subsequence or session with length greater that 1. **Singleton** is a navigation point p such that there exist B or S where $|B| = 1$ or $|S| = 1$.

The starters refer to the starting navigation points of user actions, whereas the attractors denote the users' targets. The singletons relate to the single user actions such as use of hotlists (e.g. history or bookmarks) [9].

We can formulate behavioral abstractions simply as the pairs of starters and attractors. Then it is equally important to observe the connecting elements of transitions from one task (or sub-task) to the other.

Definition 3. *(SE Elements, Connectors)*
Let $S_i = \{(p_{ik}, dp_{ik})\}_k^N$ and $S_{i+1} = \{(p_{i+1l}, dp_{i+1l})\}_l^M$ be consecutive subsequences, $S_i \rightarrow S_{i+1}$, of a browsing session. **SE element** (start-end element) of a subsequence S_i is a pair $D_i = (p_{i1}, p_{iN}) = (\psi_s, \psi_e)$. **Connector** of subsequences S_i and S_{i+1} is a pair $C_i = (p_{iN}, p_{i+1,1}) = (\eta_s, \eta_e)$.

The SE elements outline the higher order abstractions of user subgoals. Knowing the starting point, users can follow various navigational pathways to reach the target. Focusing on the starting and ending points of user actions eliminates the variance of navigational choices. The connectors indicate the links between elemental browsing patterns. This enables us to observe formation of more complex behavioral patterns as interconnected sequences of elemental patterns.

Definition 4. *(Range)*
Let $\mathcal{G} = (\mathcal{P}, \mathcal{B}, \mathcal{S})$ be a navigation space. **Range** $r_p = < r_{min}^{(p)}, r_{max}^{(p)} >$ of a point p is the minimum and maximum length of remaining sequences from the given position of $(p, d) \in B$ or $(p, dp) \in S$ to either starter or attractor points. Range of a singleton is **0**.

The range is an important point characteristic. It can be intuitively perceived as an extent of required navigational transitions in order to reach the nearest attractor or starter from a given point.

Combining the defined primitives we can express the usability of specific elements in the form of a vector.

Definition 5. *(Usability Vector)*
Usability vector μ is the normalized vector $\mu = ||(\mu_c, \mu_{r_{min}}, \mu_{r_{max}})||_2$, with respect to l_2 norm. Vector coordinates for point p, SE element D_i, and connector C_i are defined as follows:

$$p:\ \mu_c = \frac{c}{max_p(c)},\ \mu_{r_{min}} = \frac{r_{min}^{(p)}}{max_p(r_{min}^{(p)})},\ \mu_{r_{max}} = \frac{r_{max}^{(p)}}{max_p(r_{max}^{(p)})},$$

$$D_i:\ \mu_c = \frac{c}{max_{D_i}(c)},\ \mu_{r_{min}} = \frac{\frac{1}{2}(r_{min}^{(\psi_s)}+r_{min}^{(\psi_e)})}{max_{D_i}(\frac{1}{2}(r_{min}^{(\psi_s)}+r_{min}^{(\psi_e)}))},\ \mu_{r_{max}} = \frac{\frac{1}{2}(r_{max}^{(\psi_s)}+r_{max}^{(\psi_e)})}{max_{D_i}(\frac{1}{2}(r_{max}^{(\psi_s)}+r_{max}^{(\psi_e)}))},$$

$$C_i:\ \mu_c = \frac{c}{max_{C_i}(c)},\ \mu_{r_{min}} = \frac{\frac{1}{2}(r_{min}^{(\eta_s)}+r_{min}^{(\eta_e)})}{max_{C_i}(\frac{1}{2}(r_{min}^{(\eta_s)}+r_{min}^{(\eta_e)}))},\ \mu_{r_{max}} = \frac{\frac{1}{2}(r_{max}^{(\eta_s)}+r_{max}^{(\eta_e)})}{max_{C_i}(\frac{1}{2}(r_{max}^{(\eta_s)}+r_{max}^{(\eta_e)}))},$$

The above formulation enables usability evaluation of not only the navigation points, but also the users' behavioral abstractions and their connecting actions. This allows more complete usability analysis.

Since the number of occurrences, and the minimum and maximum ranges may vary substantially in values, they are scaled to 1, in order to obtain relatively comparable measurements. The scaling is indicated by dividing their actual values by the maximum observed value for each particular component. The difference between points $p \in \mathcal{P}$, and SE elements and connectors, is simply averaging the minimum and maximum range values of the pair.

The normalization of a usability vector μ can be carried out with respect to various norms, however, we consider l_2 norm. It leads to the directional cosines of the scaled numbers of occurrences, and the minimum and maximum ranges, and permits relevant numerical evaluation, as well as effective usability visualization.

3 Intranet and Data

Data used in this work was a one year period Intranet web log data of The National Institute of Advanced Industrial Science and Technology (Table 1). The majority of users are skilled knowledge workers. Intranet web portal had a load balancing architecture comprising of 6 servers providing extensive range of web services and documents vital to the organization. The Intranet services support managerial, administration and accounting processes, research cooperation with industry and other institutes, databases of research achievements, resource localization and search, attendance verification, and also numerous bulletin boards and document downloads. The institution has a number of branches at various locations throughout the country, thus certain services are decentralized. The size of a visible web space was approximately 1 GB. The invisible web size was considerably larger, but difficult to estimate due to the distributed architecture and constantly changing back-end data.

Table 1. Basic information about raw and preprocessed data used in the study

Data Volume	~60 GB	Log Records	315 005 952
Average Daily Volume	~54 MB	Clean Log Records	126 483 295
Number of Servers	6	Unique IP Addresses	22 077
Number of Log Files	6814	Unique URLs	3 015 848
Average File Size	~9 MB	Scripts	2 855 549
Time Period	3/2005 — 4/2006	HTML Documents	35 532
		PDF Documents	33 305
		DOC Documents	4 385
		Others	87 077

Daily traffic was substantial and so was the data volume. Information summary of the data is presented in Table 1. It is important to note that the data was incomplete. Although some days were completely represented, every month there were missing logs from specific servers. The server side logs also suffered a data loss due to the caching and proxing. However, because of the large data volume, the missing data only marginally affected the analysis. Web servers run the open source Apache server software and the web log data was in the combined log format without referrer.

4 Extraction and Segmentation of Browsing Behavior

Setup. Extraction and segmentation of knowledge worker Intranet browsing behavior from web log data was performed on Linux setup with MySQL database as a data storage engine for preprocessed and processed data. Analytic and processing routines were implemented in various programming languages and optimized for high performance.

Preprocessing. Data fusion of the web logs from 6 servers of a load balanced Intranet architecture was performed at the preprocessing level. Data was largely contaminated by the logs from automatic monitoring software and required filtering. During the initial filtering phase the logs from software monitors, invalid requests, web graphics, style sheets, and client-side scripts were eliminated. The access logs from scripts, downloadable and syndicated resources, and documents in various formats were preserved. Information was structured according to the originating IP address, complete URL, base URL, script parameters, date-time stamp, source identification, and basic statistics. Clean raw data was logged into a database and appropriately linked (see Table 1-right for summary).

Session Extraction. Preprocessed and databased Apache web logs (in combined log format) did not contain referrer information. The clickstream sequences were reconstructed by ordering the logs originating from the unique IP addresses according to the time-stamp information. Ordered log sequences from the specific IP addresses were divided into the browsing sessions as described in Definition 1. Session divisor was the predetermined user inactivity period ds_i greater than $T_B = 1$ *hour*.

Table 2. Observed basic session data statistics

Number of Sessions	3 454 243
Number of Unique Sessions	2 704 067
Average Number of Sessions per Day	9 464
Average Session Length	36 [URL transitions]
Average Session Duration	2 912.23 [s] (48 min 32 sec)
Average Page Transition Delay per Session	81.55 [s] (1 min 22 sec)
Average Number of Sessions per IP Address	156

It is noticeable (see Table 2) that the user sessions on the corporate Intranet were on average longer (appx. 48.5 minutes) than those of the students (appx. 25.5 minutes) reported in [8]. The average number of 156 sessions per IP address, and a large variation in the maximum and minimum number of sequences from distinct IP addresses, indicate that association of users with distinct IP addresses is relevant only for the registered static IP addresses.

Subsequence Extraction. Each detected session was analyzed for the subsequences as defined in Definition 1. It has been observed that the sessions contained machine generated subsequences. As seen in the histogram of average delays between subsequences (Figure 1-a), there was a disproportionally large number of sessions with average delays between subsequences around 30 minutes and 1 hour. This is indicated by the spikes in Figure 1-a. The detailed view (subcharts of Figure 1-a) revealed that the variation in the average delay between subsequences was approximately ±3 seconds. It correlates with the peak average subsequence duration (Figure 1-b). It is highly unlikely that human generated traffic would produce this precision. We filtered two groups of machine traffic.

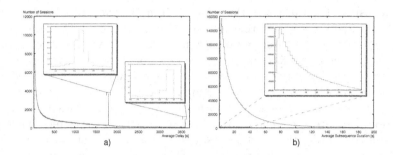

Fig. 1. Histograms: **a)** average delay between subsequences in sessions, **b)** average subsequence duration. There are noticeable spikes in chart **a)** around 1800 seconds (30 minutes) and 3600 seconds (1 hour). The detailed view is displayed in subcharts. Temporal variation of spikes corresponds to the peak average subsequence duration in chart **b)**. The spikes with relatively accurate delays between subsequences are due to machine generated traffic.

Table 3. Observed basic subsequence data statistics

Number of Subsequences	7 335 577
Number of Valid Subsequences	3 156 310
Number of Filtered Subsequences	4 179 267
Number of Unique Subsequences	3 547 170
Number of Unique Valid Subsequences	1 644 848
Average Number of Subsequences per Session	3
Average Subsequence Length	4.52 [URL transitions]
Average Subsequence Duration	30.68 [s]
Average Delay between Subsequences	388.46 [s] (6 min 28 sec)

The first group is the initial login. Every user is required to login into Intranet in order to access the services and resources. The login procedure involves validation and generates several log records with 0 delays. The records vary depending on whether the login was successful or unsuccessful. In both cases the log records and login related subsequences can be clearly identified and filtered.

The second group of machine generated traffic are the subsequences with periodicity of 30 minutes and 1 hour. Direct way of identifying these subsequences is to search for the sessions with only two subsequences having less than 1 second (or 0 second) duration (machines can generate requests fast and local Intranet servers are capable of responding within milliseconds) and the delay ds_i between subsequences within the intervals: 1800 and 3600 ± 3 seconds. It has been discovered that substantial number of such sessions contained relatively small number (170) of unique subsequences. They, and corresponding points, were filtered.

5 Intranet Usability Analysis

By analyzing the point characteristics together with behavioral abstractions and introduced usability metrics we infer several relevant observations. Analysis demonstrates effectiveness of the framework in elucidating usability features.

5.1 Starters, Attractors, and Singletons

The point characteristics of a navigation space highlight the initial and terminal targets of knowledge worker activities, and also single-action behaviors. It is evident that filtering produced substantial reduction of relevant navigation points as well as SE elements and connectors (Table 4).

Table 4. Statistics for starters, attractors, singletons, SE elements, and connectors

	Starters	Attractors	Singletons	SE elements	Connectors
Total	7 335 577	7 335 577	1 326 954	7 335 577	3 952 429
Valid	2 392 541	2 392 541	763 769	2 392 541	2 346 438
Filtered	4 943 936	4 943 936	563 185	4 943 936	1 605 991
Unique	187 452	1 540 093	58 036	1 540 093	1 142 700
Unique Valid	115 770	288 075	57 894	1 072 340	898 896

Knowledge workers utilized a small spectrum of starting navigation points and targeted relatively small number of resources during their browsing. The set of starters, i.e. the initial navigation points, was approximately 3.84% of total navigation points. Although the set of unique attractors, i.e. (sub-)goal targets, was approximately three times higher than the set of initial navigation points, it is still relatively minor portion (appx. 9.55% of unique URLs).

Knowledge workers had focused interests and exhibited minuscule exploratory behavior. A narrow spectrum of starters, attractors, and singletons was frequently

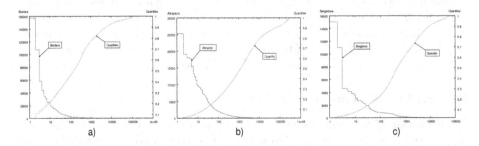

Fig. 2. Histograms and quantiles: **a)** starters, **b)** attractors, and **c)** singletons. Right y-axis contains a quantile scale. X-axis is in a logarithmic scale.

used. The histograms and quantile characteristics of starters, attractors, and singletons (see Figure 2) indicate that higher frequency of occurrences is concentrated to relatively small number of elements. Approximately ten starters and singletons, and fifty attractors were very frequent. About one hundred starters and singletons, and one thousand attractors were relatively frequent.

Knowledge workers were generally more familiar with the starting navigation points rather than the targets. Smaller number of starters repeats substantially more frequently than the adequate number of attractors. In other words, they knew where to start and were familiar with the navigational path to the target.

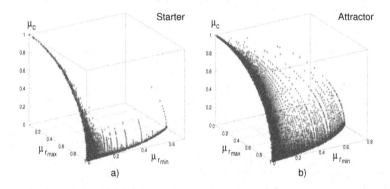

Fig. 3. Usability vector characteristics: **a)** starters, and **b)** attractors. All coordinates in 3D graphs depict normalized values.

Many infrequent starters and attractors were relatively difficult to reach. As displayed in Figure 3 there is a large number of starters and attractors with relatively small μ_c values, and relatively high values of $\mu_{r_{max}}$ and $\mu_{r_{min}}$. These are the infrequent elements (small μ_c) to which access was uneasy for most users. In order to reach these navigation points, users were required to make more transitions.

Frequent starters have relatively small minimum and maximum ranges. This is a desirable situation. Frequently accessed initial navigation points should be easily reachable. However, this condition holds only in the interval $\mu_c \in (0.9, 1)$. As μ_c values decreases below 0.8, the relative maximum range $\mu_{r_{max}}$ of starters rises beyond 0.6 (Figure 3-a). That is, relatively frequent starters were still more difficult to access. This situation should be improved.

5.2   SE Elements and Connectors

Small number of SE elements and connectors was frequently repetitive. The histogram and quantile charts in Figure 4 depict re-occurrence of the SE elements and connectors. Approximately thirty SE elements and twenty connectors were very frequent (refer to the left histogram curves of Figure 4). These thirty SE

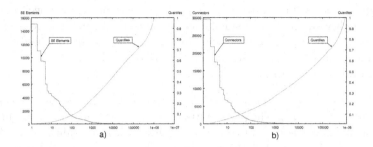

Fig. 4. Histograms and quantiles: **a)** SE elements, and **b)** connectors. Right y-axis contains a quantile scale. X-axis is in a logarithmic scale.

elements (appx. 0.0028% of the unique valid SE elements) and twenty connectors (appx. 0.0022% of the unique valid connectors) accounted for about 20% of the total observations (see the right quantile curves of Figure 4).

Knowledge workers formed elemental and complex browsing patterns. Strong repetition of the SE elements indicates that knowledge workers often initiated their browsing actions from the same navigation point and targeted the same resource. This underlines the elemental pattern formation. Re-occurrence of the connectors suggests that after completing a browsing sub-task, by reaching the desired target, they proceeded to the frequent starting point of the following sub-task(s). Frequently repeating elemental patterns interlinked with frequent transitions to other elemental sub-task highlights formation of more complex browsing patterns.

Frequent elemental and more complex browsing patterns required small number of transitions. This is a positive factor, since the often used browsing patterns and connecting transitions should be easily executable and reachable. It is noticeable

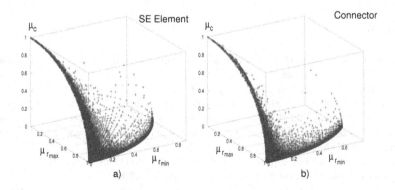

Fig. 5. Usability vector characteristics: **a)** SE elements, and **b)** connectors. All coordinates in 3D graphs depict normalized values.

from Figure 5 that a small number of frequent SE elements and connectors with a relatively high value of μ_c had small averaged relative minimum and maximum ranges $\mu_{r_{min}}$ and $\mu_{r_{max}}$, respectively. However, this conditions holds for the SE elements and connectors with μ_c values greater than 0.9; then the relative maximum range $\mu_{r_{max}}$ rises.

Substantial number of infrequent SE elements and connectors was relatively difficult to execute and reach. The plots in Figure 5 show that there is a large number of infrequent SE elements and connectors—those with low values of μ_c. Major portion of them had relatively high maximum range $\mu_{r_{max}} > 0.8$, however, still relatively low minimum range $\mu_{r_{min}} < 0.2$. Smaller, but considerable, portion of the elements had the relative minimum range $\mu_{r_{min}} > 0.4$.

6   Conclusions

A novel usability analysis framework utilizing the segmentation of user web interactions based on the periods of activity and inactivity has been introduced. The presented metrics enable effective evaluation of web navigation points as well as user behavioral abstractions. The approach has been applied to elucidation of usability attributes of a large commercial intranet portal with extensive knowledge worker user base. The knowledge workers formed elemental and complex browsing patterns that were often reiterated. They effectively utilized only small amount of available resources and exhibited diminutive exploratory behavior. General browsing strategy was familiarity with the starting navigation point and recall of the navigational path to the target.

Acknowledgment

The authors would like to thank Tsukuba Advanced Computing Center (TACC) for providing raw web log data.

References

1. Schlender, B.: Peter Drucker sets us straight. Fortune (December 29, 2003), http://www.fortune.com
2. Davenport, T.H.: Thinking for a Living - How to Get Better Performance and Results from Knowledge Workers. Harvard Business School Press, Boston (2005)
3. Petre, M., Minocha, S., Roberts, D.: Usability beyond the website: an empirically-grounded e-commerce evaluation for the total customer experience. Behaviour and Information Technology 25, 189–203 (2006)
4. Benbunan-Fich, R.: Using protocol analysis to evaluate the usability of a commercial web site. Information and Management 39, 151–163 (2001)
5. Norman, K.L., Panizzi, E.: Levels of automation and user participation in usability testing. Interacting with Computers 18, 246–264 (2006)
6. Seffah, A., Donyaee, M., Kline, R.B., Padda, H.K.: Usability measurement and metrics: A consolidated model. Software Quality Journal 14, 159–178 (2006)

7. Géczy, P., Izumi, N., Akaho, S., Hasida, K.: Extraction and analysis of knowledge worker activities on intranet. In: Reimer, U., Karagiannis, D. (eds.) PAKM 2006. LNCS (LNAI), vol. 4333, pp. 73–85. Springer, Heidelberg (2006)
8. Catledge, L., Pitkow, J.: Characterizing browsing strategies in the world wide web. Computer Networks and ISDN Systems 27, 1065–1073 (1995)
9. Thakor, M.V., Borsuk, W., Kalamas, M.: Hotlists and web browsing behavior–an empirical investigation. Journal of Business Research 57, 776–786 (2004)

Photo-Based User Profiling for Tourism Recommender Systems

Helmut Berger[1], Michaela Denk[1], Michael Dittenbach[1],
Andreas Pesenhofer[1], and Dieter Merkl[2]

[1] E-Commerce Competence Center–EC3,
Donau-City-Straße 1, A–1220 Wien, Austria
{helmut.berger,michaela.denk,michael.dittenbach,
andreas.pesenhofer}@ec3.at
[2] Institut für Softwaretechnik und Interaktive Systeme,
Technische Universität Wien,
Favoritenstraße 9–11/188, A–1040 Wien, Austria
dieter.merkl@ec.tuwien.ac.at

Abstract. The World Wide Web has become an important source of information for tourists planning their vacation. So, tourism recommender systems supporting users in their decision making process by suggesting suitable holiday destinations or travel packages based on user profiles are an area of vivid research. Since a picture paints a thousand words we have conducted an online survey revealing significant dependencies between tourism-related photographs and tourist types. The results of the survey are packaged in a Web-based tourist profiling tool. It is now possible to generate a user profile in an enjoyable way by simply selecting photos without enduring lengthy form-based self assessments.

1 Introduction

Photographs bring moments back to life – be they very personal or moments shared by many. Assuming you are interested in relaxing and sunbathing in warm places with lots of sun, sand and ocean, you will have taken, without much doubt, pictures of sunny and sandy beaches. Conversely, if your primary emphasis is to remain active while on vacation, you may engage in your favorite sports and so take snapshots of your magic moments. All these moments, and thus tourism activities, can be categorized according to some typology of tourists and, in turn, it is possible to identify the relationship between these types and tourist activities [4]. Grounded on this relationship we postulate the hypothesis that preferences for particular tourism-related photographs can be used to derive a tourist's type, and so, generate a profile of the user's likings. Currently, the process of creating such profiles can be a rather annoying, time-consuming and cumbersome task. However, intelligent services such as tourism recommender systems heavily rely on personal user profiles in addition to explicitly expressed needs and constraints. These systems focus on recommending destinations and product bundles tailored to the users' needs in order to support their decision

G. Psaila and R. Wagner (Eds.): EC-Web 2007, LNCS 4655, pp. 46–55, 2007.

making process [1,2,8]. Quite frequently, tourism recommender systems need to deal with first-time users, which implies that such systems lack purchase histories and face the cold-start problem [9]. Some systems tackle this problem by requesting the user to answer a predefined set of questions. However, these might be misunderstood or simply remain unanswered [5]. Such non-adaptive approaches are problematic, since poorly assembled user profiles reduce the quality of recommendations, and consequently, negatively effect the acceptance and success of tourism recommender systems. A different line for user preference elicitation is taken in [7], where profiles for new users are generated based on Likert-scale ratings of products. The new user is required to assess her likings until sufficient overlap to profiles of known users can be derived. This is feasible in an application setting where commodities are being sold. In tourism, however, the constraints are different since the products are generally rather expensive and annual leave is limited.

In order to prove our hypothesis, we have conducted an online survey revealing significant dependencies between tourism-related photographs and tourist types. Our results show that we can take advantage of this relationship and propose a profiling technique based on photograph selection, which minimizes the efforts for users formalizing their likings and get them as quickly as possible to the fun part. The results of the survey are further packaged in a Web-based tourist profiling tool.

The remainder of this paper is organized as follows. In Section 2 we present our online survey and some basic facts regarding the respondents. Sections 3 and 4 contain our findings from the survey regarding motivating factors and significant photographs for various tourist types. In Section 5 we show the tourist type profiler developed based on the survey results. Finally, Section 6 gives some conclusions.

2   The Online Survey

To investigate whether tourist's preferences can be derived from tourism-related photographs we conducted a survey. An online questionnaire was made public in July 2006 on a Web portal. This questionnaire consisted of three parts whereof the first part aimed at obtaining personal and demographic data of the participants. These were age group, gender, marital status, number of children, highest level of education, and whether they live in a city or town.

The second part was designed to capture the tourism preferences of the participants. They were asked to select from a set of 17 tourist types based on the tourist typology proposed by Yiannakis and Gibson [10]. The tourist types were described in terms of statements such as "interested in relaxing and sunbathing in warm places with lots of sun, sand and ocean" or "mostly interested in meeting the local people, trying the food and speaking the language" whereof the first description corresponds to the tourist type referred to as the *Sun Lover* and the latter to the *Anthropologist*. Note that we refrained from providing the actual labels of the tourist types presuming that participants might be biased by these. Additionally, we have defined four age groups, viz. less than 20, 21 to 40, 41 to 60, and over 60. Each participant was asked to select those tourist types which she has belonged to in earlier periods of her life, or currently belongs to. For example, a participant aged 47 was requested to select her personal tourism habits when she was younger than 20, between 21 and 40 as well as her current preferences.

The third part of the questionnaire comprised 60 photos depicting different tourism-related situations. Participants should identify those photos that best represent their personal tourism habits. In the end, we have gathered data from 426 respondents; their demographic composition in shown Table 1.

Table 1. Personal and demographic characteristics of survey sample (n=426)

Gender	Female - 208; Male - 218
Age group	21 to 40 - 200; 41 to 60 - 187; 61 and above - 39
Education	Primary - 148; Secondary - 156; University - 122
Marital status	Single/separated - 115; married/living with long term partner - 311
Kids	no kids - 189; one or more kids - 237
Resident of a	city - 188; village/town - 238

The tourist typology is given in Table 2 and the descriptions as provided in the questionnaire are shown. Additionally, the absolute and relative frequencies of the respondents' tourism preferences are given. Please note that the sum of the relative frequencies exceeds 100%, because most respondents obviously assigned themselves to more than one tourist type. The rank order of tourist types in this table significantly correlates (Pearson's $r = 0.895, \alpha = 0.001$) with the results presented in [3].

3   On Pack and Kick in Tourism

In order to generate a map of the relationships between tourist types and the photographs we carried out a correspondence analysis. Starting from a cross tabulation of photo click frequencies by tourist type, we obtained the correspondence analysis map depicted in Figure 1. The results show that the relationship between tourist type and photo can be mapped onto two dimensions that account for 56.44% of the inertia, i.e. a large amount of the total variance is explained by the first two principal axes. In particular, the x-axis can be referred to as the *Pack Factor* and the y-axis represents the *Kick Factor*. The *Pack Factor* identifies the "level of collectivity" one can associate with a particular tourist type. Consider, for example, the *Explorer*, which is the left-most tourist type, and the *Organized Mass Tourist*, the right-most tourist type along the x-axis. The *Explorer* might be identified as a rather solitary individual compared to an *Organized Mass Tourist*, who is generally accompanied by a larger number of like-minded tourists. This interpretation is corroborated by the findings of a study in which tourist experiences have been identified to vary along an individualistic/collectivistic continuum [6]. The *Kick Factor* identifies the "level of excitement" one might associate with a particular tourist activity. The *Thrill Seeker*, for instance, is by definition interested in risky, exhilarating activities

Table 2. Tourist types, their descriptions and distributions statistics

Tourist type	Description	Freq.	%
Anthropologist	Mostly interested in meeting the local people, trying the food and speaking the language	334	78.40
Escapist I	Enjoys taking it easy away from the stresses and pressures of home environment	320	75.12
Archaeologist	Primarily interested in archaeological sites and ruins; enjoys studying history of ancient civilizations	265	62.21
Sun Lover	Interested in relaxing and sunbathing in warm places with lots of sun, sand and ocean	263	61.74
Independent Mass , Tourist I, (IMT I)	Visits regular tourist attractions but avoids packaged vacations and organized tours	223	52.35
High Class	Travels first class, stays in the best hotels, goes to shows and enjoys fine dining	207	48.59
Independent Mass Tourist II, (IMT II)	Plans own destination and hotel reservations and often plays it by ear (spontaneous)	196	46.01
Escapist II	Gets away from it all by escaping to peaceful, deserted or out of the way places	174	40.85
Organized Mass Tourist, (OMT)	Mostly interested in organized vacations, packaged tours, taking pictures/buying lots of souvenirs	163	38.26
Active Sports	Primary emphasis while on vacation is to remain active engaging in favorite sports	158	37.09
Seeker	Seeker of spiritual and/or personal knowledge to better understand self and meaning of life	136	31.92
Explorer	Prefers adventure travel, exploring out of the way places and enjoys challenge in getting there	132	30.99
Educational Tourist, (Edu-Tourist)	Participates in planned study tours and seminars to acquire new skills and knowledge	127	29.81
Jet Setter	Vacations in elite, world class resorts, goes to exclusive night clubs, and socializes with celebrities	104	24.41
Action Seeker	Mostly interested in partying, going to night clubs and meeting people for uncomplicated romantic experiences	86	20.19
Thrill Seeker	Interested in risky, exhilarating activities which provide emotional highs for the participant	61	14.32
Drifter	Drifts from place to place living a hippie-style existence	55	12.91

that provide emotional highs. Contrary, the *Escapist I* enjoys taking it easy, away from the stresses and pressures of the home environment.

The generated layout of photos is to a high degree in-line with the alignment of the tourist types. For example, photos 22 (alpine ski touring) and 37 (alpine skiing) are highly associated with *Active Sports* whereas photos 46 (whitewater rafting), 52 (sky diving), 56 (bungee jumping) and 59 (windsurfing) correspond to the *Thrill Seeker*. The *Action Seeker* is represented by photos such as 3, 21 and 29 all of which are party sujets. The photo layout also reflects the characteristics of the axes. For example, photo 27 shows the highest level of individualism; in fact it depicts a solitary hitch hiker. On the contrary, photo 14 represents a typical packaged tour enjoyed by a group of bus tourists. In terms of the *Kick Factor*, photos 1 (car rental area at airport) and 55 (group listening to tour guide) show a moderate level of excitement whereas photos 52 and 56 depict risky and exhilarating activities. Note that photo 30 (audience with an Indian Bhagwan) was selected by 11 respondents only and, thus, is regarded as a statistical outlier.

The correspondence map is divided into four quadrants each of which reflecting peculiarities of a set of tourist types. The lower left quadrant, for example, describes a high level of individualism and rather tranquil activities. As a result, this quadrant contains tourist types such as the *Anthropologist, Archaeologist* as

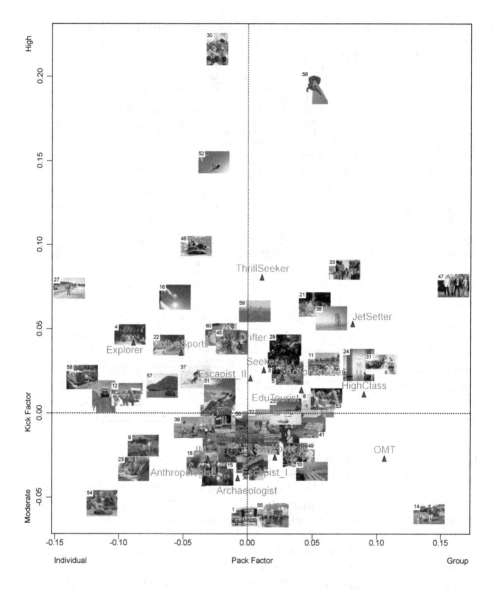

Fig. 1. Correspondence map of the relationship between tourist types and tourism-related photographs. The x–axis represents the level of collectivity and the y–axis the level of excitement.

well as the *Escapist I* that were quite frequently chosen by the respondents. The rather compact arrangement of these tourist types reflects their very close relationship and, hence, it is difficult to distinguish between them. The upper-left quadrant comprises the *Explorer*, *Active Sports* and *Drifter* tourist types that show a rather high level of individualism as well as excitement.

Seven tourist types can be found in the upper-right quadrant of the map. The types range from *Educational Tourist* to *Thrill Seeker* and from *Escapist II* to *High Class*. The differences between some of these types seem to be rather small taking their close position in the map into account. A possible interpretation is that the *Seeker* ("...searching for spiritual and/or personal knowledge...") and the *Educational Tourist* ("...searching for new skills and knowledge...") share some common ground or are performed in a sense at the same time. The lower-right quadrant contains two tourist types, namely the *Sun Lover* and the *Organized Mass Tourist*. The degree of individuality attributed to these tourist types is rather low since packaged tours can be regarded as their dominating characteristic. Nevertheless, there seems to be considerable difference in terms of individuality between the *Sun Lover* and the *Organized Mass Tourist* taking the distance of their alignment in the map into account. The *Kick Factor* associated with these tourist types is rather moderate highlighting their desire for relaxation and hassle-free tourism experiences.

4 Significant Photos

The significance of individual photos to distinguish between tourist types is analyzed by means of logistic regression. In particular, the photos with positive,

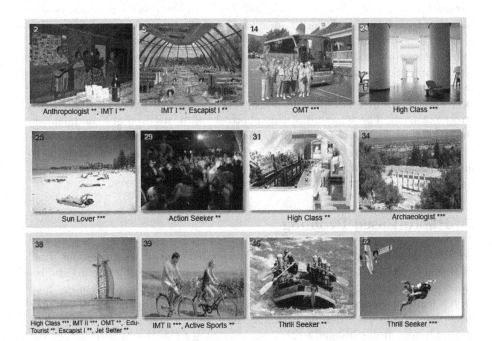

Fig. 2. Affirmative example photos for particular tourist types

Fig. 3. Counterexample photos for particular tourist types

significant coefficients in the regression model are regarded as affirmative examples for a particular tourist type. Conversely, photos with negative, significant coefficients are counterexamples. Following this approach, we obtain the mappings of photos to tourist types as given in Figure 2 and Figure 3. We indicate the significance levels with asterisks: *** ($\alpha = 0.001$) and ** ($\alpha = 0.01$).

Regarding the affirmative examples (cf. Figure 2), we obtain impressive results for characterizing the following tourist types: *Anthropologist* (photo 02 - a group of indigenous musicians), *Archaeologist* (photo 34 - the remnants of an ancient Greek temple), *Sun Lover* (photo 25 - the beach), *High Class* (photo 24 - the entrance hall of a stylish hotel; photo 31 - a posh bar), *Organized Mass Tourist* (photo 14 - group of bus tourists), *Active Sports* (photo 39 - cyclists), *Action Seeker* (photo 29 - a party), and *Thrill Seeker* (photo 46 - whitewater rafting; photo 52 - sky diving). However, we also recognized the rather unexpected phenomenon that photo 38, showing the Burj al-Arab hotel in Dubai, is representative for six tourist types. This particular photo was selected by 163 participants. So, we assume that a fairly large number of participants regarded photo 38 as their emblematic vacation dream rather than their vacation practice. For the counterexamples (cf. Figure 3), we refer to photo 54 depicting a street musician. The selection of this photo significantly excludes the membership to the *Archaeologist*. Photo 13, showing a tranquil scenery with boat, is a perfect example against the typical *Active Sports* tourist. Finally, we want to mention that only for a small number of tourist types we were unable to identify important photos, viz. *Seeker*, *Explorer* and *Drifter*.

5   Photo-Based Profiling

We have designed and implemented a Web-based tourist type profiling tool[1] as shown in Figure 4 using the statistical model established by the logistic regression. Eight sets of photographs are located in the top row and users may switch from one set to the next by clicking the respective hyperlink. Users may drag photographs they identify with into the lower-left area. If they change their mind, photos can be dragged to the area bordering right, the "Wastebin", to exclude them from their selection. Note, however, photos moved to the "Wastebin"

---

[1] Give it a go at `http://ispaces.ec3.at/TourismProfiler/index.html`

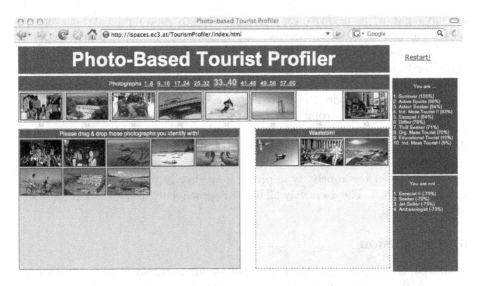

Fig. 4. Web-based tourist type profiler

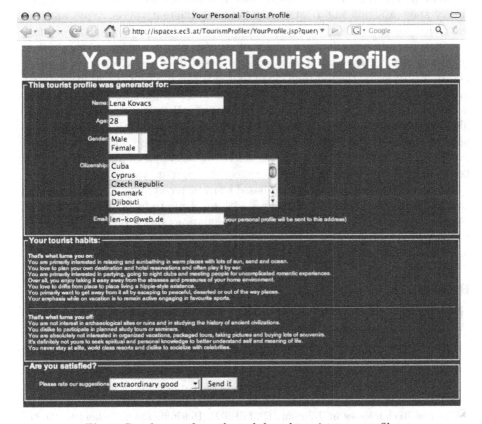

Fig. 5. Result page from the web-based tourist type profiler

are not regarded as negative examples in the regression model. Whenever the photo selection changes, the degrees of affiliation to particular tourist types is calculated instantaneously and displayed on the right-hand side of the Web page. Both, the tourist types speaking significantly for and against a person's tourism habits are shown and listed according to the descending degree of affiliation.

As an example consider the photo selection depicted in Figure 4. These photo preferences indicate a strong affiliation with the tourist type "Sun Lover", followed by "Active Sports", "Action Seeker" and "Independent Mass Tourist". Additionally, this user profile resembles a person being no "Escapist II", "Seeker", "Jet Setter" or "Archeologist".

The results of the tourist type profiling are finally presented to the user as shown in Figure 5. The users may fill in their personal details and rate the quality of the profiling.

6 Conclusion

In this paper, we presented the findings of an online survey conducted to investigate whether tourist's habits can be derived from tourism-related photographs in order to facilitate the process of user profile creation. The results of this survey show a significant relationship between different tourist types and the preference for particular visual impressions conveyed by photographs. For most tourist types, we have determined representative photos, which, in turn, allow the assignment of tourist types to persons based on their selection of a set of photos. Considering the relationship of tourist types and tourist activities stated in [4], we arrive at a mapping between tourism-related photographs and tourist activities. The concept is showcased by means of a Web-based tourist type profiling tool using the statistical model established by the logistic regression. It is now possible to make the traditional process of registration and profile generation more enjoyable by letting the user select from a couple of photos that reflect her tourism habits, and then infer her according tourist types.

In a next step towards the photo-based tourism recommender system we aim at associating the different tourist types with tourism products. Thus, a user will obtain a result of tourism product recommendations based on the preferences for particular photographs. This approach, and most importantly the product to type associations, will be evaluated by means of a comprehensive user study.

References

1. Delgado, J., Davidson, R.: Knowledge bases and user profiling in travel and hospitality recommender systems. In: Proceedings of the 9th International Conference on Information Technologies in Tourism (ENTER'02), Innsbruck, Austria, pp. 1–16. Springer, Heidelberg (2002)
2. Fesenmaier, D.R., Ricci, F., Schaumlechner, E., Wöber, K., Zanella, C.: DIETORECS: Travel advisory for multiple decision styles. In: Frew, A.J., Hitz, M., O'Connor, P. (eds.) Proceedings of the 10th International Conference on Information Technologies in Tourism (ENTER'03), Helsinki, Finland, January 29–31, 2003, pp. 232–241. Springer, Heidelberg (2003)

3. Gibson, H., Yiannakis, A.: Tourist roles – needs and the lifecourse. Annals of Tourism Research 29(2), 358–383 (2002)
4. Gretzel, U., Mitsche, N., Hwang, Y.-H., Fesenmaier, D.R.: Tell me who you are and I will tell you where to go: Use of travel personalities in destination recommendation systems. Information Technology and Tourism 7(1), 3–12 (2004)
5. Jannach, D., Kreutler, G.: Personalized user preference elicitation for e-services. In: Proceedings of the IEEE International Conference on e-Technology, e-Commerce and e-Service (IEEE'05), Hong Kong, China, March 29-April 1, 2005, pp. 604–611. IEEE Computer Society Press, Los Alamitos (2005)
6. Mehmetoglu, M.: A typology of tourists from a different angle. International Journal of Hospitality & Tourism Administration 5(3), 69–90 (2004)
7. Rashid, A.M., Albert, I., Cosley, D., Lam, S.K., McNee, S.M., Konstan, J.A., Riedl, J.: Getting to know you: Learning new user preferences in recommender systems. In: Proceedings of the 7th International Conference on Intelligent User Interfaces (IUI '02), pp. 127–134. ACM Press, New York (2002)
8. Ricci, F.: Travel recommender systems. IEEE Intelligent Systems 17(6), 55–57 (2002)
9. Schein, A.I., Alexandrin Popescul, A., Ungar, L.H., Pennock, D.M.: Methods and metrics for cold-start recommendations. In: Proceedings of the 25th Annual International ACM SIGIR Conference on Research and Development in Information Retrieval (SIGIR '02), Tampere, Finland, pp. 253–260. ACM Press, New York (2002)
10. Yiannakis, A., Gibson, H.: Roles tourists play. Annals of Tourism Research 19(2), 287–303 (1992)

Examining the Relationship Between Individual Characteristics, Product Characteristics, and Media Richness Fit on Consumer Channel Preference

Eric Brunelle and Josée Lapierre

Department of Mathematics and Industrial Engineering
École Polytechnique de Montréal
C.P. 6079, succ. Centre-ville
Montreal (Quebec), H3C 3A7
`eric-3.brunelle@polymtl.ca`

Abstract. This study examines the relationship between individual characteristics, product characteristics and media richness fit to explain the consumer channel preference. Based on prior research, hypotheses were tested with a sample of 749 consumers. The results show that the media richness fit moderates the degree of the relationship between the individual/product characteristics and the consumer channel preference. Thus, depending on the level of the media richness fit, the level of confidence in the channel, the attitude towards the channel, the experience level with the channel, the perceived risk vis-à-vis the channel, the perceived product complexity, the perceived product intangibility, and the consumer's product involvement correlate differently with the consumer channel preference. Theoretical and managerial implications of the findings and avenues for future research are discussed.

Keywords: Consumer channel preference, consumer behaviour, electronic commerce, media richness theory.

1 Introduction

The objective of this empirical study is to test the link between the antecedents of consumer channel preference and the media richness theory. Accordingly, we first present the factors, identified in past studies, which influence consumer channel preference. Second, we examine the media richness theory. Third, we describe the methodology used to test the proposed hypotheses. Fourth, we present the results and lastly, we discuss the implications of these results, the limits and avenues for future research.

2 Antecedents of Consumer Channel Preference

Factors identified in previous studies provide vital input to explain consumer channel preference. Based on the results of meta-analyses that [45], [14], and [12] presented,

G. Psaila and R. Wagner (Eds.): EC-Web 2007, LNCS 4655, pp. 56–67, 2007.

we summarized the factors that explain the consumer channel preference into two categories. The first category refers to individual characteristics such as the level of confidence in the channel [12,31] the perceived risk vis-à-vis the channel [17,22] the attitude towards the channel [4,18,46], and the experience level with the channel [37,46]. From this, we can formulate the following hypothesis:

H1: *Consumers' individual characteristics are related to the consumer channel preference. Specifically, (a) the level of confidence in the channel, (b) the attitude towards the channel, and (c) the experience level with the channel are positively related to the consumer channel preference; (d) the perceived risk vis-à-vis the channel, however, is negatively related to the consumer channel preference.*

The second category refers to product characteristics such as perceived product complexity [38], perceived product intangibility [50] and consumer' product involvement [4]. Based on these past studies, we formulate the following hypothesis:

H2: *Product characteristics are positively related to the consumer channel preference. Specifically, (a) the perceived product complexity, (b) the perceived product intangibility, and (c) the consumer's product involvement are positively related to the consumer channel preference.*

3 Media Richness Theory

The media richness theory suggests that the performance of individuals in a communication context will be function of the fit between the characteristics of the media - the media richness - and the characteristics of task to be achieved - the task analyzability. Media richness refers to a medium's capability to convey certain types of information and is determined by the medium's capacity for immediate feedback, multiple cues and senses involved, language variety, and personalization [15]. Along these dimensions, media are ranked along a continuum describing their relative richness from richest medium to leanest medium: face-to-face being the richest medium followed by telephone, voice messaging, electronic mail and Web sites [43]. Task analyzability refers to the degree to which tasks involve the application of objective, well-understood procedures that do not require novel solutions. An analyzable task corresponds to a task with clear and precise predetermined responses to potential problems and an unanalyzable task requires individuals to think about, create, or find satisfactory solutions to problems outside the domain of facts, rules or procedures. As a result, richer media such as face-to-face and group meetings are more appropriate for unanalyzable tasks, whereas leaner media such as written documents are more appropriate for analyzable tasks.

The media richness theory find support in different context and was used in order to study mainly the intra-organizational communications[1]. More specifically, the theory was used to study the use of media by workers, teleworkers, virtual teams, and managers. Also, the theory was employed to study the use of electronic mail, the use

---

[1] See [1,6,10,21,32-34,48,49].

of multimedia, the realization of negotiation tasks, the achievement of teamwork, the performance of workers, the quality of organizational communication, the transfer of knowledge, the accuracy of distance education programs, the quality of service, and the impact of media on product development. Moreover, the theory was referred to in order to better explain disappointments of the users in a computer-assisted communication context and to compare the differences between men and women in their choice to use one media rather than another.

In addition, past findings in consumer behaviour studies showed that consumers are likely to adopt different behaviours in their information research process according to the channel they use. These studies also showed that the effects of factors explaining the consumer behaviour are different according to the channel[2]. In other words, in accordance with the definition of a moderator variable proposed proposed by [3] and by [47], we believe that the channel moderates the link between individual/product characteristics and the consumer channel preference, because the channel influence the degree of the relationship between those variables. Moreover, based on the results of the studies testing the media richness theory, we actually believe that it is not only the channel that plays this moderating role, but also the fit between the information presentation format of a channel and the type of information sought during the information research stage of the purchase process. This leads us to formulate the following hypothesis:

H3: *The fit between the information presentation format of a channel and the type of information the consumer seeks moderates the relationship between the individual characteristics and the consumer channel preference. Specifically, the correlation between (a) the level of confidence in the channel, (b) the attitude towards the channel, (c) the experience level with the channel, and (d) the perceived risk vis-à-vis the channel and the consumer channel preference is significantly different between a high-fit situation and a low-fit situation.*

H4: *The fit between the information presentation format of a channel and the type of information the consumer seeks moderates the relationship between the product characteristics and the consumer channel preference. Specifically, the correlation between (a) the perceived product complexity, (b) the perceived product intangibility, and (c) the consumer product involvement and the consumer channel preference preference is significantly different between a high-fit situation and a low-fit situation.*

4 Method

In accordance with the *theory application research* presented by [7-9] we aimed to get a high internal validity. Thus, we limited the study to the comparison between two channels: brick-and-mortar store and online store. This type of comparison was used by other researchers [26,39,42,44]. Moreover, we focused the study on the consumer information research stage for a personal computer. This product appears perfectly suitable for two main reasons. Firstly, the consumers can carry out their information

---

[2] See [16,27,29,41].

research for personal computer via several channels, among which are the brick-and-mortar store and the online store. Secondly, consumers perceive personal computers differently [20]. For some, a personal computer represents a product which is highly complex and intangible, and for others, it represents a less complex product of a tangible nature. To ensure that the respondents possessed a minimum knowledge of Internet use, an online survey was designed to realize this study.

4.1 Sample

In accordance with the objective of obtaining a high internal validity, a homogeneous sample is recommended [9]. We thus limited our solicitation to one specific geographical area and to a homogenous group. We used e-mail lists provided by four university students' associations to recruit students coming from the same city. We invited the students to visit a Web site and to answer the survey. One cash prize of 500$ was offered to boost the response rate. We tested primary versions of the multi-item questionnaire on a convenience sample of 204 students. After a few adjustments to the questionnaire, we sent out the invitations to the emails lists. A total of 809 questionnaires were registered in the database. After analysis, 749 usable questionnaires were kept to test the proposed hypotheses. As we expected, the profile of the sample was homogeneous and fairly representative of the universities' student population. 58.1% of the respondents were between 18 and 25 of age, 29.4% were between 26 and 35, and the remainder were 36 and older. 78% of the respondents had a university-level education (48% at the undergraduate level and 30% at the graduate level), 14.6% had a college-level education[3] (those probably represented the newly admitted university students), 3.1% had less than a collegial-level education, and 5.6% did not answer. 53.8% of the respondents were female and 43.3% were male; 2.9% did not answer. 54.2% of the respondents had less than 20 000$ of income, 24.7% had an income between 20 001 and 40 000$, 12.1% had an income between 40 001 and 60 000$, 4.9% had more than 60 000$ of income, and 4 % did not answer. Finally, 36.3% of the respondents had already bought one computer, 49% had already bought 2 or 3 computers, 5.4% had already bought more than 3 computers, 6.3% had never bought any computer, and 3.1% did not answer. 54.4% of the respondents reported that they frequently or always use the Internet to find information for their purchasing decisions, 33.3% reported they sometimes use the Internet for that purpose, and 8.7% reported they rarely or never use the Internet in their information research process. The remaining 3.6% did not answer.

4.2 Measures

Measures were developed following the procedures proposed by [13]. Multi-item scales were generated based upon previous measures. To operationalize the survey, we presented to the respondents a few pictures and a brief description of the brick-and-mortar store and of the online store of a well-known national retailer of electronic

---

[3] In the province of Quebec, students must study two or three years at a college-level after leaving highschool and before going to university.

goods. In order to control the order effect, two questionnaires were developed and were randomly presented to the respondents. One questionnaire started with the presentation of the brick-and-mortar store option and the other started with the online store option. No significant differences between the results from both questionnaires were observed.

The online survey contained 66 items on a seven-point Likert scale. These items examined the level of confidence in the channel, the perceived risk vis-à-vis the channel, the attitude towards the channel[46], the experience level with the channel, the consumer perceived product complexity, the consumer perceived product intangibility, and the consumer product involvement. The measures for this study were developed as follows.

To measure the consumer channel preference level, we adapted the methodology used by [40], [27], and [23]. We used a scenario approach. A total of 12 items on seven-point Likert scales anchored by *not very probable that I use [the channel]* to *very probable that I use [the channel]* was used. Among these items, 3 scenarios corresponded to an unanalyzable task and 3 others to an analyzable task. For example, one scenario for an analyzable task was: *you wish to get informed about the price of an IBM personal computer with the following characteristics: an Intel Pentium 4 processor, a 100GB hard drive, and 512MB of memory.* One scenario of an unanalyzable task was as follows: *you wish to get a feel for the use of an IBM personal computer with the following characteristics: an Intel Pentium 4 processor, a 100GB hard drive, and 512MB of memory.* The same items were used for the brick-and-mortar store and for the Web site store.

To measure the level of confidence in the channel, the experience level with the channel, the consumer channel experience, and the consumer perceived product intangibility, we used items on seven-point Likert scales anchored by *strongly agree* to *strongly disagree*. More specifically, we adapted the four-item of channel confidence scale from [5], the four-item channel experience level scale from [28], and the five-item perceived product intangibility scale from [30]. On this last scale, after the pretests were carried out, we only used 3 of the 5 items. To measure the the perceived risk vis-à-vis the channel, the attitude towards the channel, the consumer channel attitude, the consumer perceived product complexity, and the consumer product involvement, we used seven-point semantic differential scales. More specifically, we adapted the four-item of perceived risk scale from [24], the three-item of channel attitude scale from [35], the three-item perceived product complexity scale from [38], and the six-item consumer product involvement scale from [35].

5   Analyses and Results

First, we ran exploratory factor analyses using a principal component analysis to determine the psychometric properties of the items composing the different scales. All items loaded as theoretically expected and no items needed to be removed. The Cronbach's alpha of the scales for all the measures ranged from 0.68 to 0.97 (see Table 1). Only one scale was below the 0.7 level, but was over the acceptable level of

0.6 [25], supporting the reliability of the measures. Second, a confirmatory factor analysis was performed using the EQS 6.1 software for the brick-and-mortar store model and for online store model. The models provided a satisfactory fit to the data, indicating the unidimensionality of the measures [2]. All factor loadings were highly significant ($p < 0.001$) and all the estimates for the average variance extracted (AVE) were higher than 0.50, with the exception of the high and the low-fit situation in the online store model [19]. As [51] and [36] presented in their studies, an AVE higher than 0.40 could be acceptable. Here, because these two measures were specifically and newly developed for the study and because the internal consistencies were quite high, we can claim that the measures of the study have adequate convergent validity. Also, we assessed discriminant validity among all of our measures by using two-factor CFA models. An unconstrained and a constrained model for each possible pair of constructs were run and compared. In all cases, the chi-square value of the unconstrained model was significantly less than that of the constrained model, supporting discriminant validity of the measures [2]. Overall, the results showed that the measures in our study possess adequate reliability and validity to go further and test the proposed hypotheses.

5.1 Tests of Hypotheses

To test the hypotheses, we first established the correlation between individual/product characteristics and the consumer channel preference. Because we want to test the moderator hypotheses in a second stage, we separated the analysis of the high and low-fit situations for the brick-and-mortar store model and for online store model to facilitate the comparison between the correlation coefficients. Thus, correlation test was used to test hypotheses 1 and 2. As we can see in the Table 2, except for the perceived product complexity in the brick-and-mortar store, all the relationships in a high-fit situation are significant, supporting H1 and H2 in a high-fit situation. Results are quite different in low-fit situation. In the online model, the relationships are significant only for the attitude toward the channel, perceived product intangibility and consumer product involvement. Moreover, all the correlation coefficients are lower in a low-fit situation than in a high-fit situation. In the brick-and-mortar store model, the relationships are significant for the attitude toward the channel, the perceived risk vis-à-vis the channel, the perceived product intangibility, the perceived product complexity, and the consumer product involvement. It is interesting to note that, for the product characteristics, the correlation coefficients are higher in a low-fit situation than in a high-fit situation. This result contrasts with the result observed in the online store model. Overall, only 3 variables are significantly correlates in every case: the attitude toward the channel, the perceived product intangibility and the consumer product involvement. The results observed fully supported H1b, H2b and H2c and partially supported H1a, H1c, H1d and H2a.

Following [3], a distinction between the degree of relationship between two variables is said to moderate the degree of this relationship. Then, since the groups compared aren't independent, we test the difference in the magnitude of the correlations by using t-test between the correlation coefficient (r) of each variable in

Table 1. Confirmatory factor analyses results

Brick-and-mortar store model			
	Average loading by dimension	AVE	Cronbach's α
High-fit (3 items)	0,792	0,635	0,71
Low-fit (3 items)	0,73	0,54	0,68
Confidence (4 items)	0,748	0,548	0,91
Experience (4 items)	0,885	0,797	0,97
Intangibility (3 items)	0,945	0,893	0,96
Risk (4 items)	0,770	0,581	0,86
Complexity (3 items)	0,759	0,604	0,84
Attitude (3 items)	0,872	0,766	0,89
Involvement (6 items)	.0.759	0.585	0.89
Goodness-of-fit statistics ($p < 0.001$)	$\chi2 = 708,139$; df = 291; $\chi2$ / df = 2,433; ΔBentler-Bonett = 0,944; CFI = 0,966; IFI = 0,966; RMSEA = 0,046		

Online store model			
	Average loading by dimension	AVE	Cronbach's α
High-fit (3 items)	0,675	0,461	0,71
Low-fit (3 items)	0,638	0,41	0,68
Confidence (4 items)	0,825	0,744	0,91
Experience (4 items)	0,915	0,891	0,97
Intangibility (3 items)	0,925	0,858	0,96
Risk (4 items)	0,784	0,572	0,86
Complexity (3 items)	0,762	0,61	0,84
Attitude (3 items)	0,819	0,689	0,89
Involvement (6 items)	0.708	0.571	0.89
Goodness-of-fit statistics ($p < 0.001$)	$\chi2 = 570,851$; df = 291; $\chi2$ / df = 1,962; ΔBentler-Bonett = 0,958; CFI = 0,979; IFI = 0,979; RMSEA = 0,038		

high-fit situation and in low-fit situation.. The results show that, except for the attitude towards the channel in the brick-and-mortar store model, all the t-tests are statistically significant. Thus H3 and H4 are fully supported, except for H4b that is partially supported.

Table 2. The relationship between individual and product characteristics and consumer channel preference

	Brick-and-mortar model (Correlation with consumer preference level)					
	High-fit	p	Low-fit	p	t-Test between high-fit and low-fit correlation	p
Experience	0,2071	****	0,0003	NS	0,206	****
Attitude	0,1408	****	0,1735	****	-0,033	NS
Confidence	0,1669	****	-0,0122	NS	0,179	****
Risk	-0,0833	**	-0,1316	****	0,048	***
Complexity	-0,0003	NS	-0,2266	****	0,226	****
Intangibility	-0,0599	**	-0,2283	****	0,168	***
Involvement	-0,0786	**	-0,1416	****	0,063	**
	Online store model (Correlation with consumer preference level)					
	High-fit	p	Low-fit	p	t-Test between high-fit and low-fit correlation	p
Experience	0,4022	****	0,0472	NS	0,355	****
Attitude	0,3294	****	0,1278	****	0,201	****
Confidence	0,371	****	0,0521	NS	0,318	****
Risk	-0,2789	****	-0,0139	NS	-0,265	****
Complexity	0,2562	****	0,0052	NS	0,251	****
Intangibility	0,2837	****	0,0644	**	0,219	****
Involvement	0,2315	****	0,1488	****	0,0827	**

Significant at $p<.10$, ** Significant at $p<.05$, *** Significant at $p<.01$, **** Significant at $p<.001$

6 Discussion and Conclusion

The results of this study have theoretical and managerial implications. First, our analyses empirically support the media richness theory in a commercial context. Our findings open a new way to look at the explanation of consumer channel preference. As already noticed, there exist differences between the models explaining the consumer channel preference in high-fit and low-fit situations. From now on, this distinction will thus have to be considered in the analysis of the factors leading to the choice of a channel. According to our study, the factors explaining the consumer channel preference that were identified in previous studies are better predictors in a high-fit situation. When the consumer finds himself in a low-fit situation, the factors dynamics becomes different. We believe that other, more extrinsic factors such as the channel proximity, the ease by which a channel can be accessed, the time constraints and so on might carry more weight in explaining the consumer channel preference in a low-fit context. Other studies will have to be undertaken in order to verify this assumption.

Moreover, we observed that there is a difference between the models explaining the consumer preference for the brick-and-mortar store and for the online store. Our results suggest that it would seem appropriate to make a clear distinction between the models that explain the consumer preference for the online store and for the brick-and-mortar store. It is indeed apparent that there exist differences and similarities between the factors leading to the choice of a particular channel. Our analyses suggest that, in a high-fit situation, individual characteristics have the same effect on the consumer preference for the online store and for the brick-and-mortar store, but product characteristics have opposite effects on the choice of the channel. For example, product characteristics are positively correlated with the consumer channel preference in the online store model and are negatively correlated with the consumer channel preference in the brick-and-mortar store model. This result suggests that consumers who perceived a product as complex and intangible and who are highly involved seem to prefer the online store and display an aversion towards the use of a brick-and-mortar store. This result can be explained by the fact that these consumers attempt to gather relevant information and to become knowledgeable about the product in order to minimize the time and displacement related costs. This consumer behavior ends up reducing the perceived product complexity and intangibility level, and actually matches the profile of highly involved consumers.

Lastly, the results of this study underscore how important it becomes for the companies to develop a deep understanding of the kind of information the consumers seek in their information research process. This knowledge will allow the companies to efficiently adapt the way they exploit the various channels and deploy their multi-channel strategies in order to better meet their consumers' informational needs. Our results tend to support the idea that the best approach to elaborating electronic commerce strategies is to establish them according to the company-consumers interface as a whole and not to limit them only to the Web site itself. Accordingly, a company must take into account the interaction between every channel by which consumers interact with the company. Moreover, companies should bet on the synergy between channels. They must set up consumer interfaces that will benefit from the complementarities that exist between the consumer information research task and the realization of the transaction and between the various types of information sought by the consumer.

Our results lead us to believe that other studies will have to be pursued in order to develop a better understanding of the consumer channel preference. First, as demonstrated by [11], the fit between the information presentation format of a channel and the type of information the consumer seeks turns out to be a matter of perception. For example, it is possible that a low-fit situation be perceived as being of a rather high-fit nature through the eyes of a consumer who shows a solid expertise in the use of a particular channel. Therefore, it would be interesting to push further the study of the fit construct, its various dimensions, and the factors influencing the perception of the fit. It could also be relevant to introduce a measure that would correspond to every consumer's own perception. Moreover, setting up a continuous variable to measure the perception of the fit would allow one to test whether or not the fit moderates the form of the relationship between the variables. Consequently, seeking a better understanding of the fit construct seems highly relevant, as it will provide a more complete explanation of the consumer channel preference. Second, since the results cannot be generalized, replicating the study in different contexts (looking at an extended geographical area, seeking a heterogeneous sample, involving more products from different categories, testing the model with different channels, etc) will improve the external validity. With this aim in mind, future research must be undertaken to provide more support to our findings. Third, let us recall that the objective of this research is to explain the consumer channel preference. We must however emphasize that this preference does not necessarily guarantee that the consumer will decide to use the preferred channel. Future research must be carried out in order to better understand the relationship between the consumer channel preference and the actual use of a channel.

References

1. Adria, M.: Making the most of e-mail. Academy of Management Executive. 14, 153–154 (2000)
2. Anderson, J.C., Gerbing, D.W.: Structural Equation Modeling in Practice: A Review and Recommended Two-Step Approach. Psychological Bulletin 103, 411 (1988)
3. Arnold, H.J.: Moderator Variables: A Clarification of Conceptual, Analytic, and Psychometric Issues. Organizational Behavior and Human Performance 29, 143–174 (1982)
4. Balabanis, G., Reynolds, N.L.: Consumer attitudes towards multi-channel retailers' Web sites: The role of involvement, brand attitude, Internet knowledge and visit duration. Journal of Business Strategies 18, 105–131 (2001)
5. Bhattacherjee, A.: Individual Trust in Online Firms: Scale Development and Initial Test. Journal of Management Information Systems 19, 211–241 (2002)
6. Cable, D.M., Yu, K.Y.T.: Managing Job Seekers Organizational Image Beliefs: the Role of Media Richness and Media Credibility. Journal of Applied Psychology 91, 828–840 (2006)
7. Calder, B.J., Phillips, L., Tybout, A.M.: Beyond External Validity. Journal of Consumer Research 10, 112–115 (1983)
8. Calder, B.J., Phillips, L., Tybout, A.M.: The Concept of External Validity. Journal of Consumer Research 9, 240–245 (1982)
9. Calder, B.J., Phillips, L., Tybout, A.M.: Designing Research for Application. Journal of Consumer Research 8, 197–211 (1981)
10. Carlson, J.R., George, J.F.: Media Appropriateness in the Conduct and Discovery of Deceptive Communication: the Relative Influence of Richness and Synchronicity. Group Decision and Negotiation 13, 191–210 (2004)
11. Carlson, J.R., Zmud, R.W.: Channel Expansion Theory and the Experiential Nature of Media Richness Perceptions. Academy of Management Journal 42, 153–170 (1999)

12. Chang, M.K., Cheung, W., Lai, V.S.: Literature derived reference models for the adoption of online shopping. Information & Management 42, 543–559 (2005)
13. Churchill, G.A.: A paradigm for developing better measures of marketing constructs. Journal of Marketing Reasearch 16, 64–73 (1979)
14. Constantinides, E.: Influencing the online consumer's behavior: the Web experience. Internet Research 14, 111–126 (2004)
15. Daft, R.L., Lengel, R.H.: Organizational information requirements, media richness and structural design. Management Science 32, 554–571 (1986)
16. Degeratu, A.M., Rangaswamy, A., Wu, J.: Consumer choice behavior in online and traditional supermarkets: The effects of brand name, price, and other search attributes. International Journal of Research in Marketing 17, 55–78 (2000)
17. Devaraj, S., Fan, M., Kohli, R.: Examination of online channel preference: Using the structure-conduct-outcome framework. Decision Support Systems 42, 1089–1103 (2006)
18. Fayawardhena, C.: Personal values' influence on e-shopping attitude and behavior. Internet Reasearch 14, 127–138 (2004)
19. Fornell, C., Larcker, D.F.: Evaluating structural equation models with unobservable variables and measurement error. Journal of Marketing Research 18, 39–50 (1981)
20. Gabrielsson, M., Kirpalani, V.H.M., Luostarinen, R.: Multiple Channel Strategies in the European Personal Computer Industry. Journal of International Marketing 10, 73–95 (2002)
21. Ganesan, S., Malter, A.J., Rindfleisch, A.: Does Distance Still Matter? Geographic Proximity and New Product Development. Journal of Marketing 69, 44–60 (2005)
22. Garbarino, E., Strahilevitz, M.: Gender difference in the perceived risk of buying online and the effects of receiving a site recommendation. Journal of Business Research 57, 768–775 (2004)
23. Gehrt, K.C., Yan, R.-N.: Situational, consumer, and retailer factors affecting Internet, catalog, and store shopping. International Journal of Retail & Distribution Management 32, 5–18 (2004)
24. Gupta, A., Su, B., Walter, Z.: Risk profile and consumer shopping behavior in electronic and traditional channels. Decision Support Systems 38, 347–367 (2004)
25. Hair, J.F., Anderson, R.E., Tatham, R.L., Black, W.C.: Multivariate Data Analysis. Prentice Hall, Upper Saddle River, N.J (1998)
26. Hensmans, M., Van Den Bosch, F.A.J., Volberda, H.W.: Clicks Vs. Bricks in the Emerging Online Financial Services Industry. Long Range Planning 34, 231–247 (2001)
27. Keen, C., Wetzels, M., de Ruyter, K., Feinberg, R.: E-tailers versus retailers: Which factors determine consumer preferences. Journal of Business Research 57, 685–695 (2004)
28. King, R.C., Xia, W.D.: Media appropriateness: Effects of experience on communication media choice. Decision Sciences 28, 877–910 (1997)
29. Korgaonkar, P., Silverblatt, R., Girard, T.: Online Retailing, Product Classifications, and Consumer Preferences. Internet Research 16, 267–288 (2006)
30. Laroche, M., Bergeron, J., Goutaland, C.: A three-dimensional scale of intangibility. Journal of Services Research 4, 26–38 (2001)
31. Lee, P.-M.: Behavioral model of online purchasers in e-commerce environment. Electronic Commerce Research 2, 75–85 (2002)
32. Lengel, R., Daft, R.: The selection of communication media as an executive skill. Academy of Management Executive 2, 225–233 (1988)
33. Lim, K.H., O'connor, M.J., Remus, W.E.: The Impact of Presentation Media on Decision Making: Does Multimedia Improve the Effectiveness of Feedback? Information & Management 42, 305–316 (2005)

34. Markus, L.M.: Electronic mail as the medium of managerial choice. Organization Science 5, 502–527 (1994)
35. Mathwick, C., Rigdon, E.: Play, flow, and the online search experience. Journal of Consumer Research 31, 324–332 (2004)
36. Menguc, B., Auh, S.: Creating a Firm-Level Dynamic Capability through Capitalizing on Market Orientation and Innovativeness. Journal of the Academy of Marketing Science 34, 63–73 (2006)
37. Montoya-Weiss, M.M., Voss, G.B., Grewal, D.: Determinants of online channel use and overall satisfaction with a relational, multichannel service provider. Journal of the Academy of Marketing Science 31, 448–458 (2003)
38. Mukherjee, A., Hoyer, W.D.: The effect of novel attributes on product evaluation. Journal of Consumer Research 28, 462–472 (2001)
39. Muthitacharoen, A., Gillenson, M.L., Suwan, N.: Segmenting Online Customers to Manage Business Resources: a Study of the Impacts of Sales Channel Strategies on Consumer Preferences. Information & Management 43, 678–695 (2006)
40. Nowlis, S.M., Simonson, I.: Sales Promotions and the Choice Context as Competing Influences on Consumer Decision Making. Journal of Consumer Psychology 9, 1–16 (2000)
41. Pavlou, P.A., Fygenson, M.: Understanding and Predicting Electronic Commerce Adoption: an Extension of the Theory of Planned Behavior. MIS Quarterly 30, 115–143 (2006)
42. Ranganathan, C., Goode, V., Ramaprasad, A.: Managing the transition to bricks and clicks. Communications of the ACM 46, 308–316 (2003)
43. Rice, R.E.: Task analyzability, use of new media, and effectiveness: A multi-site exploration of media richness. Organization Science 3, 475–500 (1992)
44. Rohm, A.J., Swaminathan, V.: A typology of online shoppers based on shopping motivations. Journal of Business Research 57, 748–757 (2004)
45. Saeed, K.A., Hwang, Y.J., Yi, M.Y.: Toward an Integrative Framework for Online Consumer Behavior Research: A Meta-Analysis Approach. Journal of End User Computing 15, 1–26 (2003)
46. Sexton, R.S., Johnson, R.A., Hignite, M.A.: Predicting Internet/e-commerce use. Internet Research: Electronic Networking Applications and Policy 12, 402–410 (2002)
47. Sharma, S., Durand, R.M., Gur-Arie, O.: Identification and analysis of moderator variables. Journal of Marketing Research 18, 291–300 (1981)
48. Trevino, K., Lengel, W., Daft, R.: Media symbolism, media richness and media choice in organizations: A symbolic interactionist perspective. Communication Research 14, 553–575 (1987)
49. Vickery, S.K., Droge, C., Stank, T.P., Goldsby, T.J., Markland, R.E.: The Performance Implications of Media Richness in a Business-to-Business Service Environment: Direct Versus Indirect Effects. Management Science 50, 1106–1119 (2004)
50. Vijayasarathy, L.R.: Product Characteristics and Internet Shopping Intentions. Internet Research-Electronic Networking Applications and Policy 12, 411–426 (2002)
51. Zhou, K.Z., Yim, C.K., Tse, D.K.: The Effects of Strategic Orientations on Technology- and Market-Based Breakthrough Innovations. Journal of Marketing 69, 42–60 (2005)

An Investigation into E-Commerce Adoption Profile for Small and Medium-Sized Enterprises in Bury, Greater Manchester, UK

Baomin Qi[1] and William McGilligan[2]

[1] Bolton Business School, University of Bolton, Deane Road, BL3 5AB, Bolton, UK
B.Qi@bolton.ac.uk
[2] Bury college, BL9 OBG, Manchester, UK

Abstract. E-commerce is the commercial transaction between and among organizations and individuals enabled by the digital technologies. Most recently, it is mainly referred to as the Internet-based electronic commerce. E-commerce provides many benefits for organisations to conduct business on the Internet. Since 1994, millions of companies have stepped into the digital world. However, due to the lack of knowledge and expertise of new technologies and many other reasons, the e-adoption rate for Small and Medium-sized Enterprises (SMEs) still lags behind. This paper investigates the e-adoption status for SMEs in the Bury area of Greater Manchester, UK. To conduct the research, a survey method is employed and questionnaires are distributed to local SMEs. The collected date is analysed and results are compared to the national statistics. Results show that the adoption rate of information and communication technology (ICT) and e-commerce in Bury is a step ahead of the UK in general.

Keywords: E-commerce, e-adoption for SMEs, regional business in the UK.

1 Introduction

Electronic commerce (EC) or e-commerce refers to the commercial transaction between and among organisations and individuals that is enabled by digital technologies, for instance the 1970's telephone-based modem supply-recorder system initiated by Baxter Health, 1980's electronic data interchange (EDI), and the French mintel videotext system [1], [2], [3]. Most recently, with the increasing use of the Internet, e-commerce mainly refers to the business transactions enabled via the Internet.

The Internet based e-commerce first emerged in 1994. Since then, e-commerce has made a huge impact on the way organisations conduct businesses. Millions of companies, pure player or clicks-mortar, have established their web presence and started conducting business online, although many of them have suffered from failure of online operations due to many reasons. The adoption of e-commerce has been growing at an ever fast pace. Many businesses have entered into this digital world, thus a huge net-associated digital economy has established. This results in a trend that without having a web presence and adopting e-commerce, the business will have fallen behind and will find it difficult to survive.

G. Psaila and R. Wagner (Eds.): EC-Web 2007, LNCS 4655, pp. 68–77, 2007.
© Springer-Verlag Berlin Heidelberg 2007

Although SMEs play a vital role in the national economy being the provider of majority of the jobs and a large percentage of the total GDP, many of them are still using the traditional methods to conduct business [4], [5]. To overcome the barriers that prevent them entering into the digital world will be crucial for them to compete in the business world [6], [7].

SMEs are historically thought to lack of knowledge and expertise in the area of new technologies including information and communication technologies (ICT), and therefore, not surprisingly, have lower new technologies uptake rates than larger businesses [8][9]. The knowledge and expertise of new technologies help raise the skills levels of staff and bring the opportunity to encourage high enterprise development, thus driving the knowledge economy [10], [11].

Bury is one of the furthest Greater Manchester (GM) districts from Manchester airport and has the highest percentage, (84.25%), of businesses classified as SMEs. This compares to just 76.8% SMEs of GM businesses [10]. If Bury is to be held up as a borough with a high knowledge economy it would be expected to have high numbers of employees in knowledge driven sectors, including ICT and computer related businesses. However, a survey from Careers NorthWest [12] revealed Bury has just 15.25% of its employees working in knowledge-driven sectors, compared to 17.01% in Bolton – the nearest town from Bury and 29.66% in Greater Manchester. This does not paint Bury in a positive light in its e-commerce adoption in Greater Manchester. This it may be interesting to determine the real picture of ICT and e-commerce adoption in Bury area.

The paper aims to investigate into the current status of e-commerce adoption in the Bury area of Greater Manchester. A questionnaire-based survey of local businesses is carried out. The data collected is compared to the national statistics by the Office for National Statistics [13]. The paper is structured as follows. The next section reviews the literature in the e-commerce adoption area for SMEs in North West. Section three discusses the research strategy and research instrument. Section four presents and analysed the collected primary data. Data comparisons with the national statistics are followed in section five. Section six draws conclusions of the research.

2 Literature Review

E-commerce brings all businesses, big or small, a series of benefits, for example low transaction cost and 24/7 availability [1]. Since its emergence e-commerce has made a huge impact on the way of businesses operate. Though, as with all new technologies, there are limitations involved that make SMEs hard to adopt. This section reviews varies aspects of e-commerce adoption for SMEs.

2.1 SMEs and E-Commerce

The digital era has brought about awareness in information and communication technologies, giving rise to new types of internet-based organisations. These organisations are capable of leveraging the power of network technologies to add value to products and services. The emergence of this new organisational paradigm and new types of products and services presents significant challenges for SMEs [14].

When SMEs attempt to adopt new technology including e-commerce a number of barriers seem to arise to make the task more difficult. Studies [15], [16], [17] show that factors such as cost of implementation of a new e-commerce system, the knowledge and expertise level of the new technology, Internet security issues, and their financial infrastructure often delay many SMEs to step into the e-commerce arena. Other barriers include the lack of personnel, less confidence of the new emerging technologies, and importantly doubts over the return-on-investment (ROI) [5]. Daniel [18] also highlighted that for SMEs to adopt e-commerce they not only need the technology infrastructure and all the other necessary means to set up e-commerce, they will also require complete support from the company. Without such support failure will inevitably happen. Due to these aspects confronted by SMEs, a large percent of them still use traditional ways to do business [13].

Drivers for SMEs to adopt e-commerce have been widely studied in the literature. It was found that adoption of e-commerce for SMEs is largely being driven by 'perceived benefits' rather than actual measures of benefits [16], [18]. Many SMEs are compelled into the digital world due to the competitive pressure initiated by the Internet technology. It suggests that SMEs adopt e-commerce technology mainly as a defensive reaction to other firms small or large [19]. In the study from Clapperton [20] he indicated that although 59% of the SMEs have built their web presence and this figure is set to rise with more businesses to have their websites, only a very small percentage has actually integrated the Internet into their business.

2.2 E-Commerce Adoption in North-West of England

As the World Wide Web drives the global economy, e-commerce has become one of the most important aspects of businesses. Nationally in general, many businesses have seized this opportunity and reshaped their business processes to digitally conduct businesses [13]. However, evidence available prior to this report suggested that business Internet and e-commerce take-up in the NorthWest was identified as the third worst of the national regions in 2003. 50% of them were not prepared to use the Internet based business support services. Internet sales were valued at £2.4bn by the NorthWest businesses against £12.5bn by the SouthWest businesses in 2004. The lack of e-commerce adoption results in NorthWest companies missing out the opportunities that e-commerce offers [13]. As declared in Department for Trade and Industry [8], *'E-commerce is likely to make a huge impact on the way we do business. It has the potential to lead to dramatic growth in trade, increase markets, improve efficiency and effectiveness and transform business processes'*. This sets the NorthWest businesses in a difficult situation to compete with other businesses nationally and globally.

2.3 Hypothesis

Bury is one of the furthest districts from Manchester airport in Greater Manchest and has a low number of ICT companies and low employment in the ICT sector for the NorthWest of England area. Furthermore, the above section revealed that the Internet and e-commerce take-up rate in NorthWest is relatively lower than other areas in England. Based on these facts, a hypothesis is proposed and will be verified at the end of the paper.

Hypothesis: The taking up of ICT and e-commerce by Bury SMEs is lower than that of its Greater Manchester peers and the national adoption rates.

3   Research Strategy

To carry out the research the survey method is employed to collect primary data from those SMEs in the Bury district. The survey method provides an efficient way to collect responses, and the size of the survey is unlimited [21]. In order to make the survey comparable, the questionnaire took the shape of those used for conducting the national survey and as well being adapted to suit the specific requirements for SMEs, for example to reflect the geographical issues and to explain concepts of the survey. The major aspects in relation to the uptake of ICT and e-commerce have been employed to make the comparison. To encourage replies, the questionnaire avoided to ask personal and sensitive questions. In addition, as ICT and e-commerce is a complex subject, questions therefore are branched into sub-questions with respondents given background information to assist where appropriate. Questions are designed to reduce the need for any consultation to be carried out. The number of questions is minimised to reduce response time and encourage completion. Each question is assessed for complexity to ensure that it could not be misunderstood and that it did not contain difficult or unclear terminology. Survey questioning also made use of a language similar to that employed with comparable surveys, keeping them simple and non-invasive. Prior to the distribution, pilot testing of the questionnaire was performed. The questionnaires were then distributed to all the respondents. The next section presents and analyses the collected data.

4   Data Collection and Data Analysis

The designed questionnaires were posted to mainly small and medium-sized enterprises surrounding Bury area in the NorthWest region. From over 5000 SMEs in this region, 320 of them were randomly chosen covering different sectors including health, education, manufacturing, and etc. Of the 320 questionnaires distributed, only 44 SMEs have replied with the completed questionnaires. This section illustrates and analysed the results of the collected data.

4.1   Company Profile and Geographical Coverage

The 44 respondents cover 13 of Bury's 16 wards. Fig. 1 presents the distribution of the company size in the survey. Therefore these results can represent the entire Bury area. Replies from Tottington and Moorside wards are much greater than other wards. The businesses covered from these replies include health/education, manufacturing, retail/wholesale,banking/financial,transport/communi-cation, chemistry, construction, etc. The result is comparable with the make-up of Bury SMEs, as indicated in Greater Manchester Learning and Skills Council [10] that the largest sectors are health/education 26.69%, retail/wholesale 20.8%, banking/financial business services 9.92% and transport/communication 4.63%. This also verifies the validity of the survey.

The majority of the respondents that took part in the survey have established their own business for more than 10 years and have survived well in a competitive environment. Only 11% of them are less than two years old. Most of the companies participated in the survey are small businesses with less than 10 employers. There is no exception since most businesses in Bury are SMEs.

4.2 ICT Take-Up and Internet Connection Rate in SMEs

The second part of the questionnaire obtains information on understanding of the ICT technologies and adoption level. The results of the survey show that nearly all respondent SMEs, 41 out of 44, have been using computers as part of their daily business operations. The survey has also revealed that the purposes for SMEs using computers in their business activities are administration with 26%, accounts/financials 21% and many other activities.

When the SMEs were further asked whether they have taken advantage of the Internet technologies, 40 out of 44 replied with the utilisation of the Internet (see Fig.2). Though for those who have Internet connections, mainly the e-mail service has been used in their daily operations. Most of them have connected to the Internet for more than three years. These results reveal that the use of computers and the Internet, e-mail in particular, has become an essential part of their businesses.

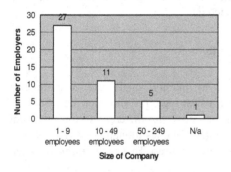
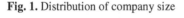
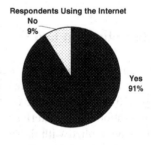

Fig. 1. Distribution of company size   **Fig. 2.** Percentage of Internet Use in the Survey

Connection to the Internet for SMEs has been achieved by using different methods. 52% of the respondents who have the Internet connection indicate that they are using broadband. The rest use dial-up or ISDN. SMEs were also asked whether they have established an intranet to automate some of the business processes, only 30% thirty percent replied with certain answers.

4.3   Company Website Establishment Rate and E-Commerce Uptake

This section delves deeper into the SMEs, their web presence and e-commerce up-take, establishing whether companies have implemented any e-commerce systems, how e-commerce has affected organisations in terms of their daily activities, what the real reasons are behind the adoptions. The survey shows that a high number of respondents have established their company websites. Thirty of the 44 replied in the affirmative as shown in Fig. 3.

This high rate of website establishment proves that most SMEs in Bury have overseen the benefit and opportunity that the Internet has brought to the company. By utilising this value-adding facility, this could improve both their online and offline sales.

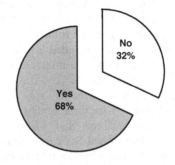

Fig. 3. Respondents with websites

Web presence can also be used as a marketing tool that serves the customer better with multi-channel communications [22]. The use of a website links to the concept of e-commerce adoption. Though in practice, the concept of e-commerce has different interpretation. Many people only consider that e-commerce is the sale and purchase from the Internet, although many activities can be considered as e-commerce, for example to obtain information from the Internet [23]. Based on this concept and also to test how much the respondents understand the e-commerce, the questionnaire included a separate section of e-commerce up-take and only asked whether they adopted e-commerce with regard to the sale and to source supplier from their website. When the respondent was asked about their e-commerce adoption, the survey results show that of the thirty companies with a website, only seven of them, which is 16% of all respondents, answered that they used e-commerce since they have been accepting orders through website. Though, this only represents the direct sale contribution from the Internet. The survey also shows that 55% of the participants used the Internet to source new suppliers. The most popular reasons for e-adoption were to 'improve sales' – the direct contribution of the Internet, and to 'gain a competitive advantage' – the pressure of a competitive environment that the Internet creates. Other reasons in the responses include: to 'save money' – use the Internet as a saving tool, and 'in anticipation of future developments', to 'help reach growth targets' and 'in response to competitors'. The gains from adopting e-commerce were in 'improved customer service' and 'improved marketing', 'improved procurement', 'improved operational effectiveness' and 'reduced costs'. This proves that the participants in the survey have realised the benefits that the Internet has brought.

The survey then asked participants whether, as a result of new technologies, they had increased or decreased staff numbers. Eight respondents claimed to have increased staff numbers with a smaller number, just two, stating that employee figures had reduced. The vast majority, 34, replied Not Applicable, suggesting either staff numbers had not changed or that this question had been given no attention. The e-commerce perception from the survey shows that majority of the respondents have viewed e-commerce as an opportunity in terms of increasing the selling of goods/services and the distribution of information.

4.4 Companies Not Adopting E-Commerce

For those who have not established a web presence and therefore not yet adopted e-commerce, it will be useful to find out why they have not had considerations and what they are going to do in the future.

The questionnaire included a section to explore the reasons for those companies currently not adopting e-commerce. From the survey, the most popular reason of not yet using e-commerce was 'Too Busy'. The second most common reason was related to 'Cost'. 'Lack of Benefits' was chosen by five respondents, followed by 'Staff Training Issues', with three replies. 'Fear of Unknown', 'Management Resistance', 'Security concerns' and 'Staff Misuse' were also quoted, although in small numbers. No companies claimed Staff Resistance as a barrier to e-commerce adoption, or the Knock-on Effect that its adoption would present. Fig. 4 shows these reasons.

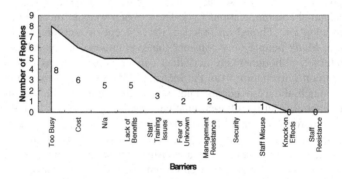

Fig. 4. Reasons given for SMEs not adopting e-commerce

When asked about the future perception of e-commerce, those 14 companies in the survey who have no websites, two planned to launch within six months, four within twelve months and six never planned. The six unplanned companies weigh 48% in this category. This implies that a large percentage of SMEs who are not currently utilising e-commerce have not yet seen the necessity to use the Internet technology. Though most of the SMEs in the survey are aware of they should have a web presence to enhance their image to help their business.

5   Comparison with National and Regional Trends

Analysis of research in the last section allows findings to be compared to large surveys to identify where Bury stands in its uptake of ICT and e-commerce in relation to the UK. The survey used to perform comparisons is the 2005 survey that was carried out by the Office for National Statistics. Since the national survey does not include the micro-companies (0-9 employees), the e-adoption profile for these companies is therefore not compared.

When compared with the national statistics, major aspects reflecting the ICT and e-commerce adoption profile are employed. These include: the use of PCs, workstations and terminals, the use of the Internet and Intranet, the up-take of websites, the utilisation of e-commerce, and the broadband up-taken.

If we take the use of PCs, workstations and terminals in 2004 [24], 85.2% of all SMEs nationally used them. Whereas in our study nearly 95.1% of Bury businesses have used PCs, Workstations or terminals to accommodate the business activities. This demonstrates that Bury is considerably ahead of the field in terms of PC, Workstation or terminals usage.   The adoption rates are further broken down according to the company size and listed in Table 1.

Looking at companies with Internet uptake, the national studies reveal that 92.2% of small and medium companies had Internet connection. However a massive 95.5% of Bury business respondents in our survey claimed to be online.

Breaking down Internet uptake across business sizes, from businesses with 10 to 49 employees, to larger businesses with up to 249 employees shows that Bury's uptake is greater across these company sizes when compared to national figures.

In the larger business groupings, up to 249 employees, all Bury respondents had Internet connection, compared to 97.6% nationally. These findings are also shown in Table 1.

Table 1. Statistical Comparison of E-commence adoption between UK and Bury

Company employees	10-49		50-249	
	UK	Bury	UK	Bury
PCs/Workstations/ Terminals Usage	91	99	99	100
Internet Uptake	86.8	91	97.6	100
Intranet Uptake	25.5	46	48.6	40
Website Uptake	65.7	63	86.7	80
Broadband Uptake	44	55	54	40

Regarding Intranet take-up, the study shows that Bury is above the national average in term of companies with 10-49 employees. The broken-down results are shown in Table 1. The average uptake rate of intranet in Bury is also higher than the national average.

In terms of the websites and e-commerce uptake rate Bury is slightly lower than the national average. 71.5% of Bury respondents had a website in the survey, compared to 76.2% of all UK businesses. Results are shown in Table 1.

The final comparison looks at Bury SME uptake of Broadband against UK business take up in 2004. Bury respondents demonstrate slightly lower compared to the national figures, with nearly half (46%) of local companies adopting Broadband, compared to 49% nationwide. If we further break down Broadband Internet uptake across business sizes, we can see that businesses in Bury with up to 49 employees had greater Broadband take-up than UK businesses, and was comparable in larger companies with up to 249 employees as displayed in Table 1.

It becomes necessary to review the proposed hypothesis in Section 2. It was initially assumed that take up of ICT and e-commerce by Bury businesses would mirror regional trends and would necessitate discussing solutions to lack of ICT and e-commerce uptake. However, the research analysis identifies that in general Bury has greater ICT and e-commerce uptake and usage than national trends, although slightly lower in one or two aspects, e.g. broadband take-up in medium-sized enterprises. This result also establishes a linkage with the survey results of e-commerce perception in Bury. Many companies in Bury area have perceived e-commerce as a new opportunity to expand their traditional businesses.

6 Conclusions

E-commerce is an emerging concept involving using innovative technologies to enable organizations and individuals to conduct business on the Internet. It offers big opportunities for business to expand and provides many benefits for both businesses and consumers [1], [2]. The utilisation and adoption becomes a necessity for business to survive and stay competitive. Nearly all large organisations have implemented some kind of e-commerce applications. Though due to the lack of knowledge on new technologies and limited resources, many SMEs have not yet stepped into the digital world. The adoption of e-commerce within large organisations creates an even competitive environment that in order to conduct business with them and therefore to survive SMEs must accept this new technology and adapt it to their existing business processes. This paper has conducted a survey to try to determine the true e-adoption status for SMEs surrounding Bury in Greater Manchester. Analysis of research results and comparisons with the national statistics revealed Bury's true ICT and e-commerce uptake rate.

The hypothesis of the research is that Bury, as one of the further districts of Greater Manchester from Manchester airport, should have the lowest uptake rate in terms of the uptake of new emerging technologies including ICT and e-commerce since business in general would be more likely to invest in locations and infrastructures that support their growth and success. However, the survey results and the comparisons with national statistics contradict with this hypothesis. Bury SMEs were comfortable with the new e-commerce technologies. Many of them have taken advantage of a web presence to improve their sale activities.

References

1. Landon, K.C., Traver, C.G.: E-Commerce: Business, Technology and Society, 4th edn. Addison Wesley, London (2006)
2. Beynon-Davies, P.: E-business. Palgrave Macmillan, Oxford (2004)
3. Shipside, S.: E-Marketing. Capstone Publishing (2005)
4. Observatory Report: SMEs in Europe, including a first glance at EU candidate countries (2002), available at http://europa.eu.int/comm/enterprise

5. eEurope GoDigital: Helping SMEs to Go Digital, online (2001), http://europa.eu.int/ISPO/ecommerce/godigital
6. Benchmarking Report: Benchmarking National and Regional Policies in Support of E-Business for SMEs (2001) online http://www.ebusinessforum/gr/docs/final-benchmarking_report.pd
7. Ivis, M.: Analysis of barriers impeding e-business adoption among Canadian SMEs, a Canadian e-Business Opportunities Round Table Report (March 2001)
8. Department for Trade and Industry: White paper: building the knowledge-drive economy, Online (2000), http://www.dti.gov.uk/comp/competitive/main.htm
9. Greater Manchester Learning and Skills Council: Census 2003 Fact Sheet Online (2003), http://www.iscgm.info
10. Development Gateway: The Knowledge Economy (2005) Online http://www.developmentgateway. org/node/130667/
11. New Zealand Ministry of Economic Development: What is the Knowledge Economy (2005) Online available at: http://www.med.govt.nz/pbt/infotech/knowledge_economy/
12. Careers Northwest: Hot Northwest careers – ICT Online (2005), http://www.careersnorthwest.com/factsheets/factsheet_sector.aspx?id=8&section=3
13. Office for National Statistics: Information and Communication Technology (ICT) (2005), http://www.statistics.gov.uk/downloads/theme_economy/ecommerce_report_2005.pdf
14. European Union eEurope: An information society for all Online (2005), http://europa.eu.int/information_society/eeurope/2002/index_en.htm
15. Turban, E., King, D., Lee, J., Warkentin, M., Chung, H.M.: Electronic Commerce 2002, a Managerial Perspective. Prentice-Hall, Englewood Cliffs (2002)
16. Daniel, E., Wilson, H.: Adoption intentions and benefits realized: a study of e-commerce in UK SMEs. Journal of Small Business and Enterprise Development 9(4), 331–348 (2002)
17. Matlay, H., Addis, M.: Adoption of ICT and e-commerce in small businesses: an HEI-based consultancy perspective. Journal of Small Business and Enterprise Development 10(3), 321–335 (2003)
18. Daniel, E.: An exploration of the inside-out model: e-commerce integration in UK SMEs. Journal of Small Business and Enterprise Development 10(3), 233–249 (2003)
19. Baldwin, J.R., Sabourin, D.: Impact of the Adoption of Advanced ICTs on Firm Performance in the Canadian Manufacturing Sector, Online (2006), http://www.oecd.org/sti
20. Clapperton, G.: Thinking beyond the brochure, The Guardian-Business Solutions, 30/10/2003, pp. 3–5 (2005)
21. Saunders, M., Lewis, P., Thomhill, A.: Research methods for business students, Pitman (2001)
22. Smith, P.R., Chaffey, D.: eMarketing excellent Butterworth, Heinemann (2006)
23. Office for National Statistics: Use of ICT in businesses in 2001, Online (2004), http://www.statistics.gov.uk/downloads/theme_economy/TBreport_revised_03Jul03.pdf

Analysis of Mobile and Pervasive Applications from a Corporate Investment Perspective

Daniel Simonovich

European School of Business,
Reutlingen University, Alteburgstrasse150, D-72762 Reutlingen
daniel.simonovich@reutlingen-university.de

Abstract. Mobile and Pervasive applications have received considerable attention over the last few years, extending the application boundaries of modern information and communication technology. However, the successful adoption of such systems in an enterprise relies on a positive investment decision. As general managers and chief information officers (CIOs) still lack exhaustive experience with ubiquitous systems, there is a need to communicate this potential to this target group, guided by reliable corporate benefit arguments. This paper presents a set of simple taxonomies that help structuring and thereby communicating the nature and benefits of mobile and pervasive systems for corporate evaluation purposes. The taxonomies were validated involving a survey gathering the opinions of 20 CIOs. The taxonomy use is portrayed in a mobile logistics application case from the chemical industry.

Keywords: Mobile Business, pervasive systems, IS evaluation, strategic fit.

1 Introduction

Information and communication technology (ICT) has witnessed mobility and pervasiveness as important developments over the last few years [1] [2], thereby propelling information systems (IS) from the networked to the ubiquitous era [3]. However, these developments are sometimes perceived as techno-centric, leaving industrial decision makers with limited orientation for establishing a management rationale from the corporate perspective [4]. Based on earlier research on the validation of pervasive applications benefits along the corporate value chain [5], this paper develops an integrative approach for systems appraisal by presenting analysis dimensions relevant in an investment decision process. To bridge the potential language gap between computer scientists and general managers, an overall framework is presented, providing relevant analysis dimension with simple taxonomies. Using a varied set of ten ubiquitous applications, these taxonomies are used to exemplify effective communication of pervasive systems impact in a corporate context. The positioning of the mentioned applications along the analysis dimensions represents the result of a survey including 20 CIOs. The benefits of the resulting structural framework are exemplified using a chemical industry case.

G. Psaila and R. Wagner (Eds.): EC-Web 2007, LNCS 4655, pp. 78–88, 2007.
© Springer-Verlag Berlin Heidelberg 2007

2 Looking at Applications from the Corporate Perspective

In an industrial applications context, mobile and pervasive systems are part of firm's IS infrastructure, usually a large set of individual applications partitioned into IS portfolio segments, each representing distinct benefit attributes [6]. A portfolio is ideally derived from an IS strategy, translating corporate objectives into ICT infrastructure [7]. These goals also direct an organization's core competence choice which is implemented through distinct business process emphasis [8]. From a financial perspective, information systems (IS) projects are evaluated by estimating the difference between business benefits and business costs [9]. These bottom line accounting estimates are, among other, gained by analyzing a system's operative effect on business process improvements. The overall big picture is shown in fig. 1. The theoretical foundation of the appraisal categories relies on the analysis dimensions referenced by the influential MIS design science research method [10].

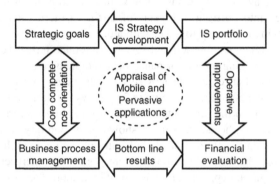

Fig. 1. Business appraisal context of mobile and pervasive systems

IS decision making – whether about traditional or ubiquitous IS – requires insights and judgment from all four appraisal dimensions. To enhance the understanding of mobile and pervasive applications within these dimensions, 20 CIOs were asked to position a wide-ranging set of ten ubiquitous applications on a five-level Likert scale (see table 1). To qualify for the survey, their organizations had to have previous exposure to ubiquitous applications. Furthermore, these individuals were required to be in

Table 1. Set of mobile and pervasive applications

a_1 Laptop based mobile field sales support	a_6 Radio based factory floor management
a_2 Mobile repair and maintenance service	a_7 Scanning based quality control
a_3 Warehouse orders using intelligent tags	a_8 Travel / expense management with PDA
a_4 Wireless inventory management	a_9 Pocket PC access to ERP financials
a_5 SMS based fleet management	a_{10} PDA based mobile office

a position to influence both business development and ICT decisions. The sample was limited to medium size industrial goods manufacturing companies. Assuming a normal distribution of the sample mean, the mean's standard deviation of any response category was no more than 0,2 on a response scale from 1 to 5.

3 Pervasive Systems as Part of a Corporate IS Portfoli

The IS portfolio is a key analysis dimension, linking distinct corporate goals to ICT applications. Understanding the benefits represented by each IS portfolio segment helps convincing non-technical decision makers using recognizable arguments. The need to distinguish distinct benefit categories can be viewed historically: On its way from the mainframe origins to the ubiquitous age, ICT has grown from a mass transaction and individual/group effectiveness tool to a broad means of achieving value creation in an organization [3]. Consequently, the spectrum of information systems can be partitioned into portfolio segments of distinct benefit categories [6]. Fig. 2 shows the underlying taxonomy together with a description of typical benefits in each segment category. Whereas transactional systems automate repetitive tasks, informational systems help judging the performance of corporate activities. Strategic systems are purposefully aimed at revenue or new service generation. Infrastructure segments usually support application segments. Figure 2 displays the designation of mobile and pervasive applications to the segments according to the dominant views of the questioned CIOs. However, the actual survey showed that contrary to conventional IS portfolio theory, a significant number of CIOs attributed a given mobile or ubiquitous application to more than one portfolio segment. One may interpret that with their ever-presence, ubiquitous applications tend to offer a more hard-to-limit mixture of benefits than their traditional MIS counterparts.

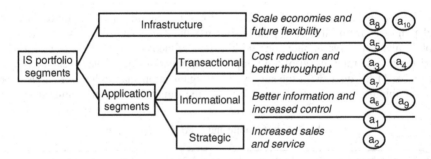

Fig. 2. IS portfolio taxonomy with positioning of mobile and pervasive applications by CIOs

Using the IS portfolio taxonomy in fig. 2, the benefits of pervasive applications can be effectively communicated to a non-technical audience. For example, advocating intelligent tracking can be supported using cost reduction and throughput arguments.

4 Strategy Impact of Pervasive Systems

Mobile and pervasive systems can bear significant investments, potentially requiring the attention and consent of top level management who may wish to see their strategic fit into corporate policy. Conversely, the adoption of such systems may be the result of a systematic search for ICT applications aimed at fitting a given strategy. Both, however, requires an understanding of how ubiquitous solutions can be referred to business strategy formulation. Consequently, taxonomy development should aim for such comprehension. The field of business strategy is rather vast. In essence, strategy development boils down to strategic analysis followed by strategic choices [11]. One of the early findings in strategy research was the identification of common stratagems known as generic strategies: The original distinction between low cost versus differentiation [12]. This was refined into the three generic strategies of operational excellence, product leadership, and customer intimacy [13]: Operational excellence signifies the promise of consistent quality at minimal costs. Customer intimacy puts the emphasis on the careful tailoring and adaptation of products and services. Product leadership denotes fast-paced innovation that yields a steady flow of leading-edge products. The evaluation of the CIO responses undertaken revealed typical correlation patterns between business strategy and IS strategy (portfolio) [14], summarized in the relational table of fig. 3: These correlations were gained by observing the category correspondence of the sample means of CIO responses across the ten applications. As a result, any portfolio segment may have a strong, weak or neglectful relationship with a given generic strategy.

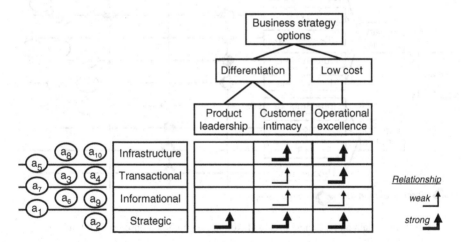

Fig. 3. Simple business strategy taxonomy related to the IS portfolio through CIO opinions

The simple taxonomy of fig. 3 can be used in support of effective communication. For example, a firm wishing to be the leader in customer intimacy should have a close look at mobile sales and mobile service applications for strategy execution purposes.

5   Pervasive Systems in Support of Business Activities

While strategic options dictate general goals a pervasive system is supposed to support, and while IS portfolio segments clarify qualitative benefits, these frameworks do not express which part of an organization is actually affected. Some business activity map is required to do so. There exist two paradigms: Value chains and business processes. The value chain concept was developed for analyzing the value created and cost incurred along activity steps, suitable for competitive comparison [15]. It distinguishes primary and support activities, striving to summarize virtually all business activities with some affinity to traditional, departmental structures. By contrast, the business process paradigm emphasizes cross departmental end-to-end activities delivering value to customers, thereby implementing the core competence philosophy of competitive advantage [8]. Based on the responses of the CIO sample group, fig. 4 shows how the two paradigms relate to one another and how ubiquitous applications fit into both frameworks.

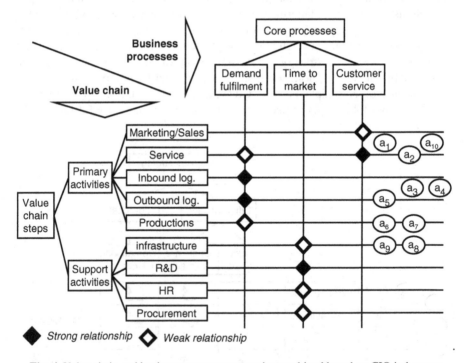

Fig. 4. Value chain and business process taxonomies combined based on CIO judgments

Analyzing fig. 4, it seems that the majority of applications predominantly support main value chain activities. Given that the set of chosen applications is rather divers, this emphasizes the potential of mobile and ubiquitous applications to value generation. Regardless of the primary attribution of applications to value chain activities, the actual CIO responses fortuitously generated strong and weak relationships between each value chain activity and some of the cross-functional core business processes:

Demand fulfillment primarily relies on high performing logistics backed up by smooth productions and responsive service. The "time to market" process falls back on R&D and major support activities. Customer service does not only insist on superior service but also requires a tight cooperation with marketing and sales. Using the classification of fig. 4, corporate mobile or pervasive systems can now be positioned in organizational terms. Together with the core process identification, it helps clarify the activity logic underlying ubiquitous functionality. Thus, mobile and pervasive systems can better be explained as a natural extension of traditional IS along a firm's value chain or mission-critical business process.

6   Financial Analysis of Pervasive Systems Investments

IS investments have distinct characteristics, turning them more difficult than usual corporate investment subjects: While costs believed to be certain are easily set off by implementation project overruns, the benefits vary from certain to inaudible [16]. The economic evaluation of internet companies is a prominent example of the limitations of financial valuation precision [17]. Academic progress was, however, made in recognizing that different IS portfolio segments typically require distinct valuation procedures. In that sense, a taxonomy may indicate when the valuation of mobile and pervasive systems may yield quantitative precision and when qualitative arguments remain as a viable alternative. The array of IS net benefit analyses has been classified into fundamental methods, combined methods, and meta-methods [18]. Fundamental methods refer to investment situations with monetary figures together with applicable techniques such as the calculation of payback periods. Combined methods are those

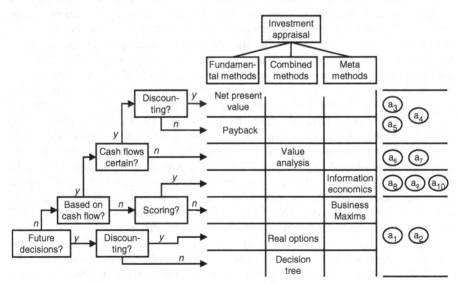

Fig. 5. Taxonomy development for IS investment evaluation

that mix fundamental techniques with non-financial analysis whereas meta models depart from monetary values. To arrive at a rigorous but simple taxonomy, a decision-oriented set of questions helps categorize contemporary evaluation methods. In the base case of cash estimation certainty the established methods of net present value (NPV) analysis and calculation of payback periods [19] apply. The inclusion of less certain cash flow estimates based on expert opinions is known as value analysis [20]. Other IS evaluation methods include business maxims [6], information economics [21], decision trees, real options analyses [22], and the balanced scorecard [23]. Fig. 5 translates the complexity maze into a taxonomy format.

The positioning of the ten mobile and pervasive applications in the taxonomy above was not directly derived from the survey responses of the CIOs – as practical experience with IS evaluation methods is still too limited to expect reliable answers. The positioning was instead created by combining the IS portfolio survey results (see figure 3) with existing research on the applicability of evaluation methods for distinct applications categories. Using the taxonomy, one might argue, for example that clear cash benefits may be demanded from wireless inventory management and less so from a mobile office suite. This may be seen as a support of managerial expectation management.

7   Case Study: Mobile Logistics at ATOTECH

Owned by the French TOTAL corporation and headquartered in Germany, ATOTECH is one of the world's leading suppliers of chemistry and equipment for the PCB and electroplating industry. Before the introduction of a mobile solution, much of the logistics chain was handled manually. Over 1500 paper based receipts were generated per day and entered manually into an ERP system through a remote desktop terminal. This not only presented a massive administrative workload but also compromised error-free and up-to-date logistics data. As a consequence, mistaken deliveries could not be recognized and stopped in time. Also any downtime of the ERP system or the connecting network often unluckily resulted in harmful delays, compromising timely delivery and customer satisfaction. The situation was changed through the introduction of a mobile logistics solution relying on eight stand-alone industrial PDAs as well as five PDAs integrated into fork-lift trucks. These PDAs are directly connected to the ERP system. The blue collar workers on the inventory shop floor directly enter data into the PDAs. Transportation orders, receipts, and status inquiries are processed on-line through wireless integration with the ERP system. During the emergence of ERP system downtime, logistics data can be administrated off-line the PDAs and be re-synchronized once the ERP system is up and running again. The implementation of the system took four months [24]. This application example shall now be used to employ the taxonomies in the previous sections.

A glance at fig. 3 reveals that the mobile logistics applications primarily fits the category of transactional application systems – the applications class deployed for cost and error reductions as well as for better throughput. By using the relationship between business strategy and IS portfolio, we determine that the competitive contribution is the support of operational excellence (the pursuit of consistent quality at low cost). Turning to business activity analysis, fig. 4 primarily discloses inbound and

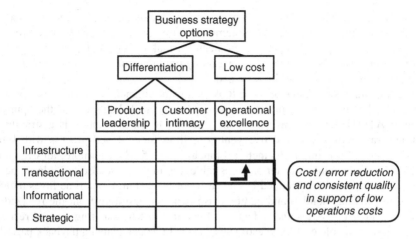

Fig. 6. Application of strategy and IS portfolio taxonomies to the ATOTECH case

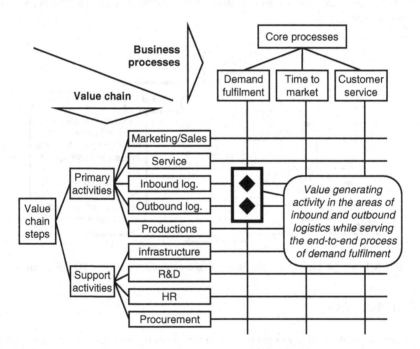

Fig. 7. Application of value chain and business process taxonomies to the ATOTECH case

outbound logistics as the value adding activities in question (although secondary value adding activities may be argued in the service and productions areas). The figure also suggest that demand fulfillment, which includes all activities from order to delivery, is the end-to-end process principally supported. The value of using such relatively simple taxonomies is that strategic and operative benefits and implications of innovative mobile or pervasive systems can be communicated and pushed forward using recognizable general management decision-making terminology such as generic strategies or value chain stages. The taxonomy outcomes of the mobile logistics case example are recapitulated in figures 6 and 7.

The final analysis dimension to be looked at is the financial evaluation perspective. A look at the financial evaluation taxonomy of figure 5 helps reveal the approach taken at ATOTECH. The investment in a mobile logistics solution is a straight investment rather than a speculative venture calling for a staged process with future decision points (see initial distinction branch in fig. 5). Furthermore, an evaluation based on real financials was not only desirable, but in this case also feasible as the net cash benefits could be calculated by the cash outflows (initial one-time investment of €70.000 over four months) and audible cash benefits, resulting from the realized reduction in false deliveries and reduction of central staff for entering manual receipts. On the basis of stable cash flow estimates, ATOTECH calculated a payback period of a little less than 9 months [24]. Given this relatively short time period, a more precise discounting based evaluation, which incorporates the time value of money, was not appropriate (see fig. 8).

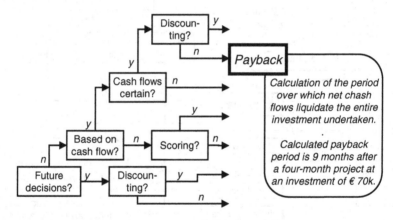

Fig. 8. Financial appraisal of the mobile logistics solutions of the ATOTECH case

8   Conclusions

The acceptance of mobile and pervasive applications by managerial decision-makers can hardly rely on technology fascination. Whereas considerable research regarding user acceptance and validation of applications has been undertaken, there remains some communication gap to reach corporate decision-makers. This target group is familiar with general management frameworks like generic strategies, value chains or investment methods. Therefore it seems worthwhile to communicate the impact of novel technology applications using such frameworks. The survey based on 20 CIOs showed that is all but hopeless to position innovative mobile and pervasive systems in well-established management vocabulary structures (taxonomies). As novel and exotic as ubiquitous applications may seem – if the taxonomies introduced are fairly complete, then the benefits and organizational meaning of these applications should still be easy to articulate in familiar profit terms. The ATOTECH example illustrated how the simple taxonomies can be applied in a real problem, real investments case. Going from here, further research can be undertaken in three directions. First, the structural work presented could be integrated into a single cohesive framework, preferably in the form of a user friendly software tool with extensive explication. Second, the set of ten inspected applications could grow to a greater reference catalogue. Finally, it should be remembered that the taxonomies, the CIO survey participants and the case example shown were taken from the industrial goods sector. Consequently, a similar investigation could be undertaken in other industry sectors such as consumer goods, retail, or service industries.

References

1. Hansmann, U., Merk, L., Nickous, M., Stober, T.: Pervasive Computing Handbook, pp. 11–25. Springer, Heidelberg (2001)
2. Fouskas, K.G., Giaglis, G.M., Kourouthanassis, P.E., Karnouskos, S., Pitsillides, A., Stylianou, M.: A roadmap for research in mobile business. International Journal of Mobile Communications 3(4), 350–373 (2005)
3. Applegate, L.M.: Managing in an Information Age – IT Challenges and Opportunities. Harvard Business School Report 9-196-004, Boston (1995)
4. Simonovich, D.: Business-Centric Analysis of Corporate Mobile Applications. In: 8th International Workshop on Mobile Multimedia Communications, pp. 269–273 (2003)
5. Simonovich, D., Malinkovich, V.: Ubiquität entlang der betrieblichen Wertschöpfungskette (Ubiquity along the Corporate Value Chain). HMD, vol. 229 (Ubiquitous Computing), dpunkt, pp. 33–41 (2003)
6. Broadbent, M., Weill, P.: Management by Maxim: How business and IT managers can create IT infrastructures. Sloan Management Review, 77–92 (1997)
7. O'Brien, J.A.: Management Information Systems: Managing Information Technology in the Business enterprise. McGraw-Hill, New York (2004)
8. Jelassi, T., Enders, A.: Strategies for E-Business – Creating Value through Electronic and Mobile Commerce, Person, pp. 114–116 (2005)
9. Wagle, D.: The case of ERP systems. McKinsey Quarterly 1998(2), 131–138 (1998)
10. Hevner, A.R., March, S.T., Parc, J., Ram, S.: Design Science in IS Research. Information Systems Quarterly 28(1), 75–85 (2004)
11. Day, G.S., Reibstein, D.J., Gunther, R.: Wharton on Dynamic Competitive Strategy, pp. 19–22. Wiley, New York (1997)
12. Porter, M.E.: Competitive Strategy – Techniques for Analyzing Industries and Competitors, pp. 34–46. Free Press, New York (1980)
13. Treacy, M., Wiersema, F.: The Discipline of Market Leaders. Addison-Wesley, London (1995)
14. Broadbent, M., Weill, P.: Leveraging the New Infrastructure, pp. 132–135. Harvard Business School Press, Boston (1998)
15. Porter, M.E.: Competitive Advantage – Creating and Sustaining Superior Performance, pp. 33–61. Free Press, New York (1985)
16. Broadbent, M., Weill, P.: Leveraging the New Infrastructure, pp. 204–230. Harvard Business School Press, Boston (1998)

17. Resch, A.: Valuation of Internet Companies – Difficult or Impossible? Shaker, Aachen, 60–62 (2000)
18. Irani, Z., Love, P.E.D.: Developing a frame of reference for ex-ante IT/IS investment evaluation. In: European Journal of Information Systems Evaluation 11, 74–82 (2002)
19. Blackstaff, M.: Finance for IT-Decision Makers, pp. 54–72. Springer, Heidelberg (1999)
20. Broadbent, M., Weill, P.: Leveraging the New Infrastructure, pp. 204–230. Harvard Business School Press, Boston (1998)
21. Parker, M., Benson, R.: Information Economics – Linking Business Performance to Information Technology. Prentice Hall, Upper Saddle River (1988)
22. Amram, M., Kulatilaka, N.: Real options – managing strategic investments in an uncertain world. Harvard Business School Press, Boston (1999)
23. Van Grembergen, W.: Measuring and Improving Corporate Information Technology through the Balanced Scorecard. In: European Journal of Information Systems Evaluation 1(1) (1997)
24. http://www.erp-mobile.de/download/success_story_atotech_web.pdf

Online Shopping Using a Two Dimensional Product Map

Martijn Kagie, Michiel van Wezel, and Patrick J.F. Groenen

Econometric Institute, Erasmus University Rotterdam, The Netherlands
{kagie,mvanwezel,groenen}@few.eur.nl

Abstract. In this paper, we propose a user interface for online shopping that uses a two dimensional product map to present products. This map is created using multidimensional scaling (MDS). Dissimilarities between products are computed using an adapted version of Gower's coefficient of similarity based on the attributes of the product. The user can zoom in and out by drawing rectangles. We show an application of this user interface to MP3 players and give an interpretation of the product map.

1   Introduction

In most electronic commerce stores, customers can choose from an enormous number of different products within a product category. Although it is expected that increased choice is better for customer satisfaction, this is not the case [1]. This phenomenon is known as the paradox of choice. When the amount of choice options increases, customers often end up choosing an option that is further away from the product they prefer most. One reason for this phenomenon is that it is very hard to get an overview of all the products that are available.

In many product categories, such as real estate and electronics, a consumer has to choose from a heterogenous range of products with a large amount of product attributes. Often, the customer first has to make a selection based on a (limited) number of constraints on product attributes, before a subset of products satisfying these constraints is shown to her. These products are usually shown in a list. A disadvantage of this approach is that customers can find these constraints too strict. In addition, product attributes can substitute each other, that is, a higher value on one attribute can compensate for a lower value on another. In this way, selection on pairs of attributes may not allow for attribute combinations that are preferred by a consumer. For example, a consumer who wants to buy an MP3 player can be equally satisfied with a cheaper MP3 player with less memory as with a more expensive MP3 player that has also more memory. Sometimes, lists of products are constructed using some measure of similarity to a query. In that case all products are shown to the user, but they are ordered by relevance.

A disadvantage of the usual approach of presenting the products in a (ordered) list is that no information is given on how similar products are to each other. For example, two products that have almost the same similarity to a query can differ from this query on a completely different set of attributes and thus differ

G. Psaila and R. Wagner (Eds.): EC-Web 2007, LNCS 4655, pp. 89–98, 2007.

a lot from each other. Also, the systems only show or a subset of the products available or a complete list of products. In this way, it is impossible for customers to get a good overview of the complete range of products.

In this paper, we propose a graphical user interface for online shopping that is based on a two dimensional map of the product space. This product map is created using multidimensional scaling (MDS) [2]. The distances in this map correspond to the dissimilaties between products: Distant products in the map are very dissimilar whereas closely products are similar. These dissimilarities between products that are the input for MDS are based on the attributes of the products.

Since the product map would become very unclear, when we represent every product by a thumbnail picture, we choose to represent only a limited number of prototypical products by such an image. A clustering algorithm is used to determine these products. Furthermore, the GUI facilitates a method to zoom-in on specific parts of the map.

Our interface has some similarity with earlier work. Graphical applications using two dimensional maps, so-called inspiration interfaces, are used in the field of industrial design engineering [3,4,5]. These applications are used to explore databases in an interactive way. At first, a small set of items is shown in a 2D space. Then, the user can click in any point in space and a new item that is closest to that point is added to the space. In Kagie, Van Wezel, and Groenen [6], a recommender systems using 2D spaces (called graphical shopping interface) was proposed, that can be used to navigate through the product space. A small set of products is shown to the user each time. By clicking on one of the products, a new set of products more similar to this product is shown together with the selected product to the user in the 2D space. Both the inspiration interfaces and the graphical shopping interface only create maps based on subsets of products making it difficult to the user to get an overview of the complete product space. In contrast, the method we propose only uses a single product map of the complete set of products.

Also in somewhat related fields like news [7,8], the web [9], music playlists [10,11], and image browsing [12] GUI's based on 2D visualizations have been created only using different visualization techniques like self-organizing maps, classical scaling, Sammon mapping, and treemap.

The remainder of this paper is organized as follows. In Sect. 2, we give a description of the methodology used to implement the user interface. In Sect. 3 we introduce the user interface and in Sect. 4 an application of the our approach on MP3 Players is given. Finally, we give conclusions and recommendations.

2   Methodology

The MDS method we use to create the product map visualizes a dissimilarity matrix, in our case dissimilarities between products, in a low dimensional space. Therefore, we first need to determine a measure of dissimilarity between products. To this end, we introduce some notation. Consider a data set D, which contains products $\{\mathbf{x}_i\}_1^n$ having K attributes $\mathbf{x}_i = (x_{i1}, x_{i2} \ldots x_{iK})$. For each product, we also have a binary vector $\mathbf{m}_i = (m_{i1}, m_{i2} \ldots m_{iK})$, containing values of 1 for nonmissing attribute values. In most applications, these attributes have mixed types, that is, the attributes can be numerical, binary, or categorical.

The most often used (dis)similarity measures, like the Euclidean distance, Pearson's correlation coefficient, and Jaccard's similarity measure, are only suited to handle one of these attribute types. Also, these measures cannot cope with missing values in a natural way. Therefore, we use a dissimilarity measure which is based on the general coefficient of similarity proposed by Gower [13], which was introduced by Kagie et al. [6]. Although we will use this dissimilarity measure during this paper, since it has some specific advantages concerning the data set we use, the product map approach can be applied to any dissimilarity measure. This implies that this approach can also be used when no explicit attribute information is available, but only, for instance, co-purchases or item-item rating correlations.

The dissimilarity δ_{ij} between products i and j is defined as the square root of the average of nonmissing dissimilarity scores δ_{ijk} on the K attributes. When we let C be the set of categorical attributes and N be the set of numerical attributes, we can write the dissimilarity as

$$\delta_{ij} = \sqrt{\frac{\left(\sum_{k \in C} m_{ik} m_{jk} \delta_{ijk}^C + \sum_{k \in N} m_{ik} m_{jk} \delta_{ijk}^N\right)}{\sum_{k=1}^K m_{ik} m_{jk}}} \; . \tag{1}$$

The computation of the dissimilarity score is dependent on the type of the attribute. For numerical attributes the dissimilarity score is the normalized absolute distance

$$\delta_{ijk}^N = \frac{|x_{ik} - x_{jk}|}{\left(\sum_{i<j} m_{ik} m_{jk}\right)^{-1} \sum_{i<j} m_{ik} m_{jk} |x_{ik} - x_{jk}|} \; . \tag{2}$$

The categorical dissimilarity score is defined as

$$\delta_{ijk}^C = \frac{1(x_{ik} \neq x_{jk})}{\left(\sum_{i<j} m_{ik} m_{jk}\right)^{-1} \sum_{i<j} m_{ik} m_{jk} 1(x_{ik} \neq x_{jk})} \; , \tag{3}$$

where $1()$ is the indicator function returning a value of 1 when the condition is true and 0 otherwise.

Equation (1) is used as dissimilarity measure to compute the dissimilarity matrix $\boldsymbol{\Delta}$, having elements δ_{ij}, that is used as input for MDS.

The aim of MDS is to find a low dimensional Euclidean representation of a dissimilarity matrix such that distances between pairs of points represent the dissimilarities as closely as possible. This objective can be formalized by minimizing the raw Stress function [14]

$$\sigma_r(\mathbf{Z}) = \sum_{i<j} (\delta_{ij} - d_{ij}(\mathbf{Z}))^2 \; . \tag{4}$$

Here, the matrix $\mathbf{Z}$, having elements z_{is}, is the $n \times 2$ coordinate matrix representing the n products in two dimensions, δ_{ij} is the dissimilarity between objects i and j forming the symmetric dissimilarity matrix $\mathbf{\Delta}$, and

$$d_{ij} = \left(\sum_{s=1}^{2} (z_{is} - z_{js})^2 \right)^{1/2} \tag{5}$$

is the Euclidean distance between the coordinates of objects i and j.

The Stress function can be adapted by allowing different weights for the dissimilarities. When we define w_{ij} as the weight of dissimilarity δ_{ij}, forming the $n \times n$ matrix $\mathbf{W}$, the raw Stress becomes

$$\sigma_r(\mathbf{Z}) = \sum_{i<j} w_{ij}(\delta_{ij} - d_{ij}(\mathbf{Z}))^2 \ . \tag{6}$$

Since dissimilarities with a higher weight have more impact on the raw Stress these dissimilarities will be better represented in the final space.

In our dissimilarity measure, we allowed for missing values. Only, a dissimilarity that is based on a large number of nonmissing dissimilarity scores is more reliable than a dissimilarity based on only a few nonmissing dissimilarity scores and should receive a higher weight. This can be done by defining the weights as the proportion of nonmissing attributes used for pair ij

$$w_{ij} = K^{-1} \sum_{k=1}^{K} m_{ik}m_{jk} \ . \tag{7}$$

To minimize $\sigma_r(\mathbf{Z})$, we use the SMACOF algorithm [15] based on majorization. One of the advantages of this method is that it is reasonable fast and that the iterations yield monotonically improved Stress values.

Summarizing, the adapted Gower's coefficient and MDS are used to create a two dimensional map of the complete product space. First, the dissimilarity matrix $\mathbf{\Delta}$ with the dissimilarities between all products is computed using (1). Then, the weight matrix $\mathbf{W}$ used in MDS is computed using (7). Finally, $\mathbf{\Delta}$ and $\mathbf{W}$ are used as input for the MDS procedure returning the $n \times 2$ matrix $\mathbf{Z}$, the coordinate space of the map.

In the map, a small number of products is represented by thumbnail pictures. These are the products that the customer can use to get an overview of the product space.

Therefore, it would be nice, when these products represent different groups of products in this space. To achieve this, we use k-means clustering [16] to create clusters in this space. We decided to perform a k-means clustering on the map instead of a hierarchical clustering procedure on the original dissimilarities for two reasons. The first reason is that we want to have a clustering that is consistent with the map, while the hierarchical procedure uses the original dissimilarities and, thus, may cluster the original data better, but does not have to be consistent with the map, meaning that products which are in the same cluster do not have to be neighboring products on the map. The second reason is the higher time complexity of the hierarchical algorithms. Since clustering will be performed online in our application, this will make the clustering too slow.

To represent P products by a thumbnail picture, we have to create P clusters. After we have applied the k-means clustering algorithm on map $\mathbf{Z}$, we compute for each cluster which product is closest to the cluster center. We define $\mathbf{C}$ as the $P \times 2$ cluster center matrix and $\mathbf{G}$ as the $n \times P$ cluster membership matrix having a value of 1 at position ip when product i belongs to cluster p and a value of 0 otherwise. When we denote the index of the product representing cluster p as i_p, we can write

$$i_p = \arg\min_i \sum_{s=1}^{2} (z_{is} - c_{ps})^2 \qquad i : g_{ip} = 1 \ . \tag{8}$$

3 Graphical User Interface

In this section we show a prototype of our interface. A screenshot of this proto-type for MP3-players is shown in Fig. 1. This application can be visited online at http://people.few.eur.nl/kagie/prodmap.htm. The main part of the GUI is a large focus map, showing a selection of the complete product map. A couple of products, in this case 12, are represented by a thumbnail picture, the other products are visualized as dots. We use 12 clusters, since we think that this is a value large enough to represent the heterogeneity in the product space and not that large that the user has to consider too many products. Dots that belong to the same cluster and, thus, are represented by the same prototypical product have the same color. The map of the complete product space is shown at the top in a small overview map. Left and right from the overview map, there is a button: The zoom-in and zoom-out button. In the overview map, all products are represented by dots. However, the dots matching a thumbnail picture in the focus map are given another color (green). The red dot is the dot matching the product that is described in detail at the right side of the GUI. The product of which the description is shown can be changed by clicking on another thumb-nail picture. The zoom-in and zoom-out button are used in combination with a drawn rectangle. When pressing the zoom-in button, a rectangle drawn in the focus map determines the new bounds of the focus map. When the zoom-out button is pressed, a rectangle drawn in the overview map is used for this.

As we discussed above, the larger focus map shows a subspace of the complete product map. Initially, this subspace equals the complete space, that is, the complete map is also shown in the focus map. The user can adapt this subspace by drawing a rectangle in the overview or focus map. The products inside this rectangle form the new subset of products. Let $\begin{pmatrix} r_{\min,1} & r_{\min,2} \\ r_{\max,1} & r_{\max,2} \end{pmatrix}$ be the lower and upper bounds in dimensions 1 and 2 of the rectangle. Then $\mathbf{Z}^*$ (with size $n^* \times 2$) are the coordinates of products inside the rectangle such that

$$r_{\min,s} \leq z_{is}^* \leq r_{\max,s} \tag{9}$$

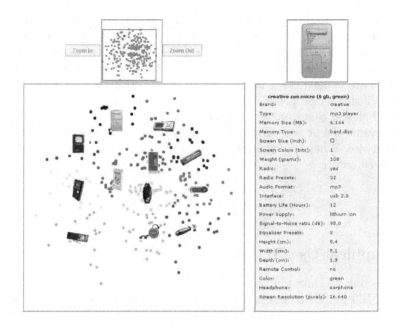

Fig. 1. Screenshot of the prototype's GUI

holds for all i in $\mathbf{Z}^*$ and $s = 1, 2$. To determine which products become the new prototypical products and thus are represented by a thumbnail picture, we apply the clustering procedure discussed in Sect. 2 on $\mathbf{Z}^*$. Note that clustering is not necessary when $n^* \leq P$, since then all products are represented by a thumbnail picture.

4   Application to MP3-Players

In the previous section we showed a prototype using MP3-player data. In this section we evaluate our approach on these data. This data set consists of 22 attributes of 321 MP3-players collected from the Dutch website http://www.kelkoo.nl using a spider during June 2006. The data set is of a mixed type, which means that we have both categorical and numerical attributes. Some attributes have many missing values. An overview of the data set is given in Table 1.

Facilitating internet shopping using a product map is only useful, when the products are positioned in the map in a way clear to the user. Since MDS visualizes high dimensional data in a low dimensional space, this may not always be the case. Formally, the quality of a MDS map can be evaluated by looking at the normalized Stress [2]. The map created of the MP3-player data has a normalized Stress value of 0.0914 what means that 90.86% of the sum-of-squares of the weighted dissimilarities is explained in the map.

Table 1. Description of the MP3-player data set. The data set describes 321 MP3-players using 22 product attributes.

Categorical Attributes	Missing	Levels (frequency)
Brand	0	Creative (53), iRiver (25), Samsung (25), Cowon (22), Sony (19), and 47 other brands (207)
Type	11	MP3 Player (254), Multimedia Player (31) USB key (25)
Memory Type	0	Integrated (231), Hard Disc (81), Compact Flash (8), Secure Digital (1)
Radio	9	Yes (170), No (139), Optional (3)
Audio Format	4	MP3 (257), ASF (28), AAC (11), Ogg Vorbis (9), ATRAC3 (5), and 4 other formats (6)
Interface	5	USB 2.0 (242), USB 1.0/1.1 (66), Firewire (6), Bluetooth (1), Parallel (1)
Power Supply	38	AAA x 1 (114), Lithium Ion (101), Lithium Polymeer (45), AA x 1 (17), AAA x 2 (4), Ni Mh (3)
Remote Control	9	No (289), In Cable (13), Wireless (10)
Color	281	White (7), Silver (5), Green (5), Orange (4), Purple (4), Red (4),Pink (4), Black (4),Blue (3)
Headphone	15	Earphone (290), Chain Earphone (8), Clip-on Earphone (2), Earphone With Belt (2), No Earphone (2), Minibelt Earphone (1), Collapsible Earphone (1)

Numerical Attributes	Missing	Mean	Stand. Dev.
Memory Size (MB)	0	6272.10	13738.00
Screen Size (inch)	264	2.16	1.04
Screen Colors (bits)	0	2.78	5.10
Weight (grams)	66	83.88	84.45
Radio Presets	9	3.06	7.84
Battery Life (hours)	40	18.63	12.56
Signal-to-Noise Ratio (dB)	247	90.92	7.32
Equalizer Presets	0	2.60	2.22
Height (cm)	28	6.95	2.48
Width (cm)	28	5.57	2.82
Depth (cm)	28	2.18	4.29
Screen Resolution (pixels)	246	31415.00	46212.00

However, more important is that the map also has a clear interpretation. In Fig. 2A and B, the values of two different attributes, memory size and weight, are used to label the products. Both panels show a similar labeling pattern. MP3 players with high values (large memory size and heavy) for these attributes are positioned at the bottom left, while MP3 players with a low value (limited memory size and light) for these attributes are positioned on the right side. A number of other attributes correlating with these two attributes are also well represented on this dimension. Two of them are the categorical attributes type and memory type shown by the labeling in Fig. 2C and D. The MP3 player of the type "multimedia player" and the memory type "hard disc" are located on the left side of the map. MP3 players of the type "MP3 player" and the memory type "integrated" are on the left side of the plot. Note that for both of these attributes, the attribute values of the products on the left side of the map correspond to large MP3 players, while the attribute values of the products on right side of the map correspond the small MP3 players.

Figure 2E shows the radio attribute that is most important in determining the second dimension. MP3 players with radio are positioned at the top of the map and MP3 players without radio at the bottom. The radio attribute is also

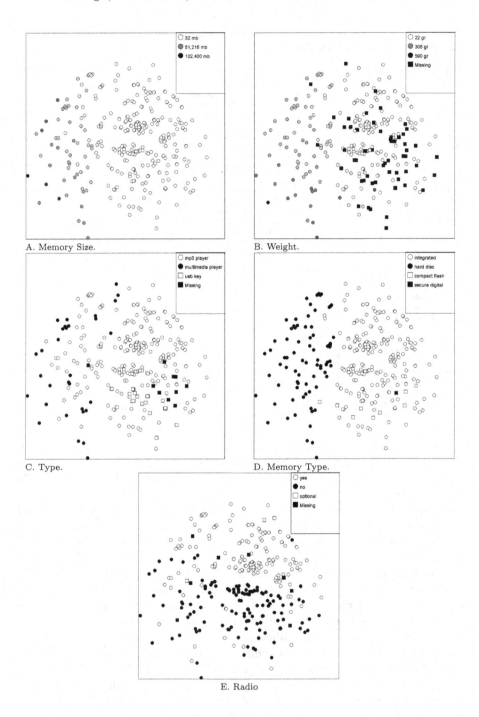

Fig. 2. Product maps wherein products are labeled by a certain attribute

one of the few attributes that have nothing to do with the size of an MP3 player. Since a radio can nowadays be implemented in an MP3 player using a tiny chip, both small and large MP3 players can have a radio.

5   Conclusions

In this paper, we presented a user interface for online shopping that uses a two dimensional product map to present products. This map is created using multidimensional scaling. Dissimilarities between products were computed using an adapted version of Gower's coefficient of similarity based on the attributes of the product.

The user interface is based on two maps. A small overview map showing the complete product map and a larger focus map showing a subspace of the complete product map. In this map, some products are represented by a thumbnail picture. These products are chosen using a k-means clustering procedure.

A prototype application to MP3 players shows promising results, since it leads to a map wherein the MP3 players are located in a natural way. The map shows two dimensions. The first can be determined as a size dimension. In the second dimension MP3 players with a radio are separated from the MP3 players without a radio.

In the current product map, all product attributes are considered to be equally important. In the perception of the user, this is generally not the case. Therefore, the map may be improved, when we are able to learn the user's preferences. This is possible when we include a possibility to rate products. Based on these ratings, it should be possible to learn weights for the different characteristics.

Furthermore, it may be the case that other visualization methods, such as principal component analysis, categorical principal component analysis [17], and projection pursuit [18] create maps with a better interpretation.

Although our approach led to a map with a reasonable interpretation, user experiments have to be carried out, to test whether users prefer the use of two dimensional products maps over traditional methods.

References

1. Schwartz, B.: The Paradox of Choice: Why More Is Less. HarperCollins, New York (2004)
2. Borg, I., Groenen, P.J.F.: Modern Multidimensional Scaling, 2nd edn. Springer Series in Statistics. Springer, Heidelberg (2005)
3. Keller, I.: MDS-i for 1 to 1 e-commerce: A position paper. In: Proceedings of the CHI 2000 Workshop on 1-to-1 E-commerce (2000)
4. Stappers, P.J., Pasman, G.: Exploring a database through interactive visualised similarity scaling. In: Altom, M.W., Williams, M.G. (eds.): Human Factors in Computer Systems. CHI99, pp. 184–185 (extended Abstracts) (1999)
5. Stappers, P.J., Pasman, G., Groenen, P.J.F.: Exploring databases for taste or inspiration with interactive multi-dimensional scaling. In: Proceedings IEA 2000 / HFES 2000, Ergonomics for the new Millennium, The Human Factors Society of the USA, pp. 3-575–3-578 (2000)

6. Kagie, M., Van Wezel, M., Groenen, P.J.F.: A graphical shopping interface based on product characteristics. In: Oria, V., Elmagarmid, A., Lochovsky, F., Saygin, Y. (eds.) Proceedings of the 23rd International Conference on Data Engineering Workshops, pp. 791–800. IEEE Computer Society, Los Alamitos (2007)

7. Ong, T.H., Chen, H., Sung, W., Zhu, B.: Newsmap: a knowledge map for online news. Decis. Support Syst. 39, 583–597 (2005)

8. Weskamp, M., Albritton, D.:   Newsmap. (2007), http://www.marumushi.com/apps/newsmap/index.cfm.

9. Turetken, O., Sharda, R.: Developement of a fisheye-based information search processing aid (FISPA) for managing information overload in the web environment. Decis. Support Syst. 37, 415–434 (2004)

10. van Gulik, R., Vignoli, F., Van der Wetering, H.: Mapping music in the palm of your hand, explore and discover your collection. In: Proceedings of the 5th International Conference on Music Information Retrieval (2004)

11. Donaldson, J.: Music recommendation mapping and interface based on structural network entropy. In: Oria, V., Elmagarmid, A., Lochovsky, F., Saygin, Y. (eds.) Proceedings of the 23rd International Conference on Data Engineering Workshops, pp. 811–817. IEEE Computer Society, Los Alamitos (2007)

12. Pečenović, Z., Do, M.N., Vetterli, M., Pu, P.: Integrated browsing and searching of large image collections. In: Goos, G., Hartmanis, J., van Leeuwen, J. (eds.) Advances in Visual Information Systems: 4th International Conference, VISUAL 2000. Proceedings. Lecture Notes in Computer Science, vol. 1929, pp. 173–206 Springer, Heidelberg (2000)

13. Gower, J.C.: A general coefficient of similarity and some of its properties. Biometrics 27, 857–874 (1971)

14. Kruskal, J.B.: Multidimensional scaling by optimizing goodness of fit to a nonmetric hypothesis. Psychometrika 29(1), 1–27 (1964)

15. De Leeuw, J.: Convergence of the majorization method for multidimensional scaling. J. Classif. 5, 163–180 (1988)

16. Hartigan, J.A., Wong, M.A.: A k-means clustering algorithm. J. R. Stat. Soc. Ser. C-Appl. Stat. 28, 100–108 (1979)

17. Gifi, A.: Nonlinear multivariate analysis. Wiley, Chichester (1990)

18. Friedman, J.H., Tukey, J.W.: A projection pursuit algorithm for exploratory data analysis. IEEE Trans. Comput. 22, 881–890 (1974)

Impact of Relevance Measures
on the Robustness and Accuracy
of Collaborative Filtering*

JJ Sandvig, Bamshad Mobasher, and Robin Burke

Center for Web Intelligence
School of Computer Science, Telecommunications and Information Systems
DePaul University, Chicago, Illinois, USA
{jsandvig,mobasher,rburke}@cs.depaul.edu

Abstract. The open nature of collaborative recommender systems present a security problem. Attackers that cannot be readily distinguished from ordinary users may inject biased profiles, degrading the objectivity and accuracy of the system over time. The standard user-based collaborative filtering algorithm has been shown quite vulnerable to such attacks. In this paper, we examine relevance measures that complement neighbor similarity and their influence on algorithm robustness. In particular, we consider two techniques, significance weighting and trust weighting, that attempt to calculate the utility of a neighbor with respect to rating prediction. Such techniques have been used to improve prediction accuracy in collaborative filtering. We show that significance weighting, in particular, also results in improved robustness under profile injection attacks.

1 Introduction

An adaptive system dependent on anonymous, unauthenticated user profiles is subject to manipulation. The standard collaborative filtering algorithm builds a recommendation for a target user by combining the stored preferences of peers with similar interests. If a malicious user injects the profile database with a number of fictitious identities, they may be considered peers to a genuine user and bias the recommendation. We call such attacks *profile injection attacks* (also known as *shilling* [1]). Recent research has shown that surprisingly modest attacks are sufficient to manipulate the most common CF algorithms [2,1,3]. Such attacks degrade the objectivity and accuracy of a recommender system, causing frustration for its users.

In this paper we explore the robustness of certain variants of user-based recommendation. In particular, we examine variants that combine similarity metrics with other measures to determine neighbor utility. Such relevance weighting techniques apply a weight to each neighbor's similarity score, based on some value reflecting the expected relevance of that neighbor to the prediction task.

---

* This work was supported in part by the National Science Foundation Cyber Trust program under Grant IIS-0430303.

G. Psaila and R. Wagner (Eds.): EC-Web 2007, LNCS 4655, pp. 99–108, 2007.

We focus on two types of relevance measures: significance weighting and trust-based weighting. Significance weighting [4] takes the size of profile overlap between neighbors into account. This prevents neighbors with only a few commonly rated items from dominating prediction. Trust-based weighting [5] estimates the utility of a neighbor as a rating predictor based on the historical accuracy of recommendations given by the neighbor.

Traditional user-based collaborative filtering algorithms focus exclusively on the degree of similarity between the target user and its neighbors in order to generate predicted ratings. However, the "reliability" of the neighbor profiles is generally not considered. For example, due to the sparsity of the data, the similarities may have been obtained based on very few co-rated items between the neighbor and the target user resulting in sub-optimal predictions. Similarly, unreliable neighbors that have made poor predictions in the past may have a negative impact on prediction accuracy for the current item. Both of the approaches to relevance weighting mentioned above were, therefore, initially introduced in order to improve the prediction accuracy in user-based collaborative filtering.

In the trust-based model [5] an explicit trust value is computed for each user, reflecting the "reputation" of that user for making accurate recommendations. Trust is not limited to the macro profile level, and can be calculated as the reputation a user has for the recommendation of a particular item. The trust values, in turn, can be used as relevance weights when generating predictions. In [5], O'Donovan and Smyth further studied the impact of trust weighting approach on the robustness of collaborative recommendation and showed the trust-based models are still vulnerable to attacks. On the other hand, the significance weighting approach, introduced initially in [4], does not focus on trust, but rather on the number of co-rated items between the target user and the neighbors as a measure for the degree of reliability of the neighbor profiles. This approach has been shown to have a significant impact on the accuracy of predictions, particularly in sparse data sets.

Although these and other similar approaches have been used to improve the prediction accuracy of recommender systems, the impact of neighbor significance weighting on algorithm robustness in the face of malicious attacks has been largely ignored. The primary contribution of this paper is to demonstrate that relevance weighting is an important factor in determining the robustness of a collaborative filtering algorithm. Choosing an optimal relevance measure can yield a large improvement in recommender stability. Our results show that significance weighting, in particular, is not only more accurate; it also improves algorithm robustness under profile injection attacks that have compact profile signatures.

2   Attacks in Collaborative Recommenders

We assume that an attacker intends to bias a recommender system for some economic advantage. This may be in the form of an increased number of recommendations for the attacker's product, or fewer recommendations for a competitor's product.

A collaborative recommender database consists of many user profiles, each with assigned ratings to a number of products that represent the user's preferences. User-based collaborative filtering algorithms attempt to discover a neighborhood of user profiles that are similar to a target user. A rating value is predicted for all missing items in the target user's profile, based on ratings given to the item within the neighborhood. A ranked list is produced, and typically the top 20 or 50 predictions are returned as recommendations.

The standard k-nearest neighbor algorithm is widely used and reasonably accurate [4]. Similarity is computed using Pearson's correlation coefficient, and the k most similar users that have rated the target item are selected as the neighborhood. This implies a target user may have a different neighborhood for each target item. It is also common to filter neighbors with similarity below a specified threshold. This prevents predictions being based on very distant or negative correlations. After identifying a neighborhood, we use Resnick's algorithm to compute the prediction for a target item i and target user u:

$$p_{u,i} = \bar{r}_u + \frac{\sum_{v \in V} sim_{u,v}(r_{v,i} - \bar{r}_v)}{\sum_{v \in V} |sim_{u,v}|} \tag{1}$$

where V is the set of k similar neighbors; $r_{v,i}$ is the rating of i for neighbor v; $\bar{r}_u$ and $\bar{r}_v$ are the average ratings over all rated items for u and v, respectively; and $sim_{u,v}$ is the Pearson correlation described above.

A malicious user may insert multiple profiles under false identities designed to promote or demote the recommendation of a particular item. Such *attack profiles* include a rating for the target item, and some number of other ratings. An *attack type* is an approach to constructing attack profiles, based on knowledge about the recommender system, its rating database, its products, and/or its users. In a push attack, the target item is generally given the maximum allowed rating. The set of *filler items* represents a group of selected items in the database that are assigned ratings within the attack profile. Attack types can be characterized according to the manner in which they choose filler items, and the way that specific ratings are assigned.

A variety of attack types have been studied for their effectiveness against different recommendation algorithms [3,6]. In this paper, we focus on the *average attack*, an attack type that has been shown to be very effective against standard user-based collaborative filtering recommenders. The average attack attempts to mimic general user preferences in the system, by drawing its ratings from the rating distribution associated with each filler item [1,3].

3   Relevance Measures

The prediction formula depicted in (1) discounts the contribution of a neighbors prediction according to its degree of similarity with the target user. Therefore, more similar neighbors have a larger impact on the final prediction. However, this type of similarity weighting alone may not be sufficient to guarantee accurate predictions. It is also necessary to ensure the reliability of the neighbor profiles. A common reason for the lack of reliability of predictions may be that similarities between the target user and the neighbors are based on a very small number of co-rated items. In the following section we consider two approaches that have been used to address the "reliability" problem mentioned above. These approaches have been used primarily to increase prediction accuracy. Our focus, however, will be on their impact on system robustness in the face of attacks. We conjenture that an optimal relevance weight may provide an algorithmic approach to securing recommender systems against attacks.

The basic goal of a relevance measure is to estimate the utility of a neighbor as a rating predictor for the target user. The standard technique is to calculate similarity as the degree of "closeness" in Euclidean space. This is often accomplished via Pearson's correlation coefficient or vector cosine coefficient. Additional extensions to similarity are well known, including significance weighting [4], variance weighting [4], case amplification [7], inverse user frequency [7], default voting [7], and profile trust [5]. In this paper, we focus on the effects of significance weighting and profile trust because they are widely accepted techniques with very different properties.

3.1   Significance Weighting

The significance weighting approach proposed by Herlocker, et al. [4] is to adjusts similarity weights by devaluing relationships with a small number of commonly rated items. It uses a linear drop-off for neighbors with less than N co-rated items. Neighbors with more than N co-rated items are not devalued at all. The significance weight of a target user u for a neighbor v is computed as:

$$w_{u,v} = \begin{cases} sim_{u,v} * \frac{n}{N} & \text{if } n < N \\ sim_{u,v} & \text{otherwise} \end{cases} \tag{2}$$

where n is the number of co-rated items, N is a global constant, and $sim_{u,v}$ is Pearson's correlation coefficient. A prediction for the target user is computed using (1), replacing $sim_{u,v}$ with $w_{u,v}$.

In addition to the standard measure, we have tested several variations. One variation that is quite successful uses a local measure and smooth devaluation, rather than a sharp cutoff. The significance weight is computed as: $w_{u,v} = sim_{u,v} * \frac{lg_n}{lg_m}$, where n is the number of co-rated items, and m is the total number of ratings in the target user's profile. Using a local measure prevents unduly penalizing the closest neighbors when the target user has only a minimal number of ratings.

Significance weighting prefers neighbors having many commonly rated items with the target user. Neighbors with fewer commonly rated items may be pushed out of the neighborhood, even if there is a higher degree of similarity to the target user. It follows that users who have rated a large number of items will belong to more neighborhoods than those users who have rated few items. This is a potential security risk in the context of profile injection attacks. An attack profile with a very large number of filler items will necessarily be included in more neighborhoods, regardless of the rating value. As we will show, the risk is minimized precisely because a large filler size threshold is required to make the attack successful. In most cases, genuine users rate only a small portion of all recommendable items; therefore, an attack profile with a very large filler size is easier to detect [8].

3.2   Trust Weighting

The vulnerabilities of collaborative recommender systems to attacks have led to a number of recent studies focusing on the notion of "trust" in recommendation. O'Donovan and Smyth [5,9] propose trust models as a means to improve accuracy in collaborative filtering. The basic assumption is that users with a history of being good predictors will provide accurate predictions in the future. By explicitly calculating a trust value, the reputation of a user can be used as insight into the user's relevance to recommendation. Trust is not limited to the macro profile level, and can be calculated as the reputation a user has for the recommendation of a particular item.

The trust building process generates a trust value for every user in the training set by examining the predictive accuracy of the corresponding profile. By cross-validation, each user in turn is designated as the sole neighbor v for all remaining users. The system then computes the prediction set P_v as all possible predictions $p_{u,i}$ that can be made for user $u \in U$ and item $i \in I$ using the neighborhood $V = v$. For each prediction $p_{u,i}$, $recommend_{v,u,i} = 1$ if $p_{u,i} \in P_v$ and $correct_{v,u,i} = 1$ if $|p_{u,i} - r_{u,i}| < \varepsilon$ where ε is a constant threshold and $r_{u,i}$ is the rating of user u for item i. Item-trust values are then computed as:

$$trust_{v,i} = \frac{\sum\limits_{u \in U} correct_{v,u,i}}{\sum\limits_{u \in U} recommend_{v,u,i}} \qquad (3)$$

There are several ways to incorporate values from the trust model into recommendation. One possibility is to filter neighbors having a trust value less than some threshold, prior to similarity calculation. However, in evaluating the robustness of relevance weighting, other aspects of neighborhood formation are control variables and remain unchanged. For our evaluation, we have chosen to use a relevance weight defined as the harmonic mean of item-trust and similarity [5]. For a target user u, the relevance weight of a neighbor v and item i is:

$$w_{u,v,i} = \frac{2 * sim_{u,i} * trust_{v,i}}{sim_{u,i} + trust_{v,i}} \qquad (4)$$

where $sim_{u,v}$ is Pearson's correlation coefficient. A prediction for the target user is computed using (1), replacing $sim_{u,v}$ with $w_{u,v,i}$.

Trust-based collaborative filtering algorithms can be very susceptible to profile injection attacks, because mutual opinions are reinforced during the trust building process [9]. Attack profiles that contain biased ratings for a target item result in mutual reinforcement of the item's preference. The larger the attack, the more reinforcement of the target item. Furthermore, if the target item is always given the maximum value, an attack profile could have higher trust scores than a genuine profile, because $correct_{v,u,i}$ will always be 1 if v and u are both attacks on item i.

In a recent study, O'Donovan and Smyth [9] propose several solutions to the reinforcement problem that utilize pseudo-random subsets of the training data during the trust building phase. Sampling the population of profiles used in trust calculation effectively smoothes the noise inherent in the entire dataset. The strategy raises an interesting research question with respect to robustness: how does a non-deterministic neighborhood formation task affect the impact of a profile injection attack? Although promising, we did not evaluate sampling the training set. For this set of experiments, we are interested only in the effect of relevance weighting.

4   Experimental Evaluation

Dataset. In our experiments, we have used the publicly-available Movie-Lens 100K dataset[1]. This dataset consists of 100,000 ratings on 1682 movies by 943 users. All ratings are integer values between one and five, where one is the lowest (disliked) and five is the highest (liked). Our data includes all users who have rated at least 20 movies.

To conduct attack experiments, the full dataset is split into training and test sets. Generally, the test set contains a sample of 50 user profiles that mirror the overall distribution of users in terms of number of movies seen and ratings provided. The remaining user profiles are designated as the training set. All attack profiles are built from the training set, in isolation from the test set.

The set of attacked items consists of 50 movies whose ratings distribution matches the overall ratings distribution of all movies. Each movie is attacked as a separate test, and the results are aggregated. In each case, a number of attack profiles are generated and inserted into the training set, and any existing rating for the attacked movie in the test set is temporarily removed.

For every profile injection attack, we track *attack size* and *filler size*. Attack size is the number of injected attack profiles, and is measured as a percentage of the pre-attack training set. There are approximately 1000 users in the database, so an attack size of 1% corresponds to about 10 attack profiles added to the system. Filler size is the number of filler ratings given to a specific attack profile, and is measured as a percentage of the total number of movies. There are approximately 1700 movies in the database, so a filler size of 10% corresponds to about 170 filler ratings in each attack profile. The results reported below represent averages over all combinations of test users and attacked movies.

---

[1] http://www.cs.umn.edu/research/GroupLens/data/

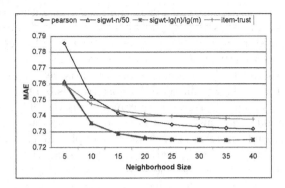

Fig. 1. Comparison of MAE

Evaluation Metrics. There has been considerable research in the area of recommender system evaluation focused on accuracy and performance [10]. We use the mean absolute error (MAE) accuracy metric, a statistical measure for comparing predicted values to actual user ratings [4]. However, our overall goal is to measure the effectiveness of an attack; the "win" for the attacker. In the experiments reported below, we follow the lead of [2] in measuring stability via prediction shift.

Prediction shift measures the change in an item's predicted rating after being attacked. Let U and I be the sets of test users and attacked items, respectively. For each user-item pair (u, i) the prediction shift denoted by $\Delta_{u,i}$, can be measured as $\Delta_{u,i} = p'_{u,i} - p_{u,i}$, where p and p' represent the prediction before and after attack, respectively. A positive value means that the attack has succeeded in raising the predicted rating for the item. The average prediction shift for an item i over all users in the test set can be computed as: $\Delta_i = \sum_{u \in U} \Delta_{u,i} / |U|$.

The average prediction shift is then computed by averaging over individual prediction shifts for all attacked items. Note that a strong prediction shift does not guarantee an item will be recommended - it is possible that other items' scores are also affected by an attack, or that the item score is so low that even a prodigious shift does not promote it to "recommended" status.

Accuracy Analysis. We first compare the accuracy of k-nearest neighbor using different relevance metrics. In our experiments we examined the standard Pearson's correlation, standard significance weighting, local significance weighting, and item-trust weighting. For significance weighting, we have followed the lead of [4] in using $N = 50$. For trust weighting, we have followed the lead of [5] in using $\varepsilon = 1.8$. In all cases, 10-fold cross-validation is performed on the entire dataset and no attack profiles are injected.

As shown in Figure 1, we achieved good results using a neighborhood size of $k = 30$ users for all relevance metrics; therefore, we applied $k = 30$ to all neighborhood formation tasks in the attack results discussed below. Overall, it is clear that some form of relevance weighting, in addition to similarity, can improve prediction accuracy. Standard and local significance weighting are particularly beneficial, although trust is also helpful when considering small neighborhoods.

There are several interesting observations about the MAE results. At $k = 5$, item-trust is more accurate than the other relevance measures. At $k = 15$ and greater, item-trust is the least accurate of the measures. It appears that the trust building process overfits the data, because trust is built on the assumption that the user for whom a trust value is computed is the only neighbor in any given neighborhood. The trust model does not take into account that a large neighborhood depends on reinforcement. For example, the closest neighbor to a target user may predict a negative rating for item i. But, when the closest three neighbors are taken into account, the second and third neighbors may predict a positive rating for item i. This effectively cancels out the prediction of the closest neighbor. In fact, a positive rating prediction may be more accurate for item i because the trend of the closest neighbors is a positive rating.

Robustness Analysis. To evaluate the robustness of relevance weighting, we compare the results of push attacks using the four relevance weighting schemes described in the previous section. Figure 2(A) depicts prediction shift results at different attack sizes, using a 5% filler. Clearly, significance weighting is much more robust than the standard Pearson's correlation. For all attack sizes, the prediction shift of significance weighting is about half that of standard correlation. Although not completely immune to attack, it is certainly a large improvement. Even at a 15% attack, significance weighting may be the difference between recommending an attacked item or not.

Local significance weighting also performs well against profile injection attack, although not to the same degree of robustness as standard significance weighting. This can be explained by the fact that target users with fewer than 50 ratings do not scale their neighbors linearly. An attack profile in the neighborhood that is highly correlated to the target user is not devalued enough. As a result, a genuine user with less correlation to the target user, but more overlap in rated items, may be removed from the neighborhood.

Item-trust weighting appears slightly more robust than standard correlation. The mutual-reinforcement effect is not as pronounced for attack profiles at smaller filler sizes, because the attacks don't have enough similarity to the target user; the trust value is outweighed. In addition, the reinforcement from genuine users is enough to gain insight into the true relevance for making predictions. The combination of trust and similarity of genuine users to a target user is sufficient to remove some attack profiles from the neighborhood.

To evaluate the sensitivity of filler size, we have tested a full range of filler items. The 100% filler is included as a benchmark for the potential influence of an attack. However, it is not likely to be practical from an attacker's point of view. Collaborative filtering rating databases are often extremely sparse, so attack profiles that have rated every product are quite conspicuous. Of particular interest are smaller filler sizes. An attack that performs well with few filler items is less likely to be detected. Thus, an attacker will have a better chance of actually

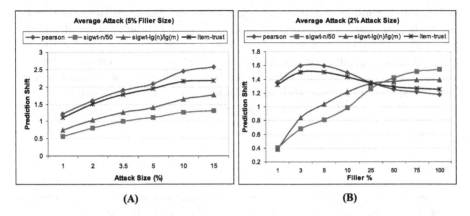

Fig. 2. (A) Average attack prediction shift at 5% filler; (B) Average attack filler size comparison

impacting a system's recommendation, even if the performance of the attack is not optimal.

Figure 2(B) depicts prediction shift at different filler sizes with 2% attack size. Surprisingly, as filler size is increased, prediction shift for standard correlation goes down. This is because an attack profile with many filler items has greater probability of being dissimilar to the active user. On the contrary, prediction shift for significance weighting goes up. As stated previously, an attack profile with a very large number of filler items will have a better chance of being included in more neighborhoods, because it isn't devalued by significance weighting.

The counter-intuitive observation is that standard correlation is actually more robust than any of the other relevance measures at very large filler sizes. To account for this, recall that the size of profile overlap is not addressed with standard correlation. A genuine user that is very similar to the target user, but does not have many co-rated items, is not penalized. However, with significance weighting the same user would be devalued, potentially removing the user from the neighborhood in favor of an attack profile.

As shown, a 25% filler size is the point where prediction shift for standard correlation surpasses the other relevance measures. Overall, this does not affect the general improvement in robustness of relevance measures. Using the modest Movie-Lens 100K dataset, a user would have to rate 420 movies to have a profile with 25% filler. It is simply not feasible for a genuine user to rate 25% of the items in a commercial recommender such as Amazon.com, with millions of different products. From a practical perspective, the threat of large filler attacks is minimal because they should be easily detectable [8].

5   Conclusion

The standard user-based collaborative filtering algorithm has been shown quite vulnerable to profile injection attacks. An attacker is able to bias recommendation by building a number of profiles associated with fictitious identities. In this paper, we have demonstrated the relative robustness and stability of supplementing the similarity weighting of neighbors with significance weighting and item-trust values. Significance weighting, in particular, results in increased recommendation accuracy and improved robustness under attack, versus the standard k-nearest neighbor approach. Future work will examine other relevance measures with respect to attack, including case amplification, inverse user frequency, and default voting.

References

1. Lam, S., Riedl, J.: Shilling recommender systems for fun and profit. In: Proceedings of the 13th International WWW Conference, New York (May 2004)
2. O'Mahony, M., Hurley, N., Kushmerick, N., Silvestre, G.: Collaborative recommendation: A robustness analysis. ACM Transactions on Internet Technology 4(4), 344–377 (2004)
3. Mobasher, B., Burke, R., Bhaumik, R., Williams, C.: Towards trustworthy recommender systems: An analysis of attack models and algorithm robustness. ACM Transactions on Internet Technology 7(4) (2007)
4. Herlocker, J., Konstan, J., Borchers, A., Riedl, J.: An algorithmic framework for performing collaborative filtering. In: Proceedings of the 22nd ACM Conference on Research and Development in Information Retrieval (SIGIR'99), Berkeley, CA (August 1999)
5. O'Donovan, J., Smyth, B.: Trust in recommender systems. In: Proceedings of the 10th International Conference on Intelligent User Interfaces (EC'04), pp. 167–174. ACM Press, New York (2005)
6. Mobasher, B., Burke, R., Sandvig, J.J.: Model-based collaborative filtering as a defense against profile injection attacks. In: Proceedings of the 21st National Conference on Artificial Intelligence, pp. 1388–1393. AAAI Press, Stanford, California (July 2006)
7. Breese, J., Heckerman, D., Kadie, C.: Empirical analysis of predictive algorithms for collaborative filtering. In: Uncertainty in Artificial Intelligence. Proceedings of the Fourteenth Conference, New Orleans, LA, pp. 43–53. Morgan Kaufman, Seattle, Washington (1998)
8. Williams, C., Bhaumik, R., Burke, R., Mobasher, B.: The impact of attack profile classification on the robustness of collaborative recommendation. In: Proceedings of the 2006 WebKDD Workshop, held at ACM SIGKDD Conference on Data Mining and Knowledge Discovery (KDD'06), Philadelphia (August 2006)
9. O'Donovan, J., Smyth, B.: Is trust robust?: An analysis of trust-based recommendation. In: Proceedings of the 5th ACM Conference on Electronic Commerce (EC'04), pp. 101–108. ACM Press, New York (2006)
10. Herlocker, J., Konstan, J., Tervin, L.G., Riedl, J.: Evaluating collaborative filtering recommender systems. ACM Transactions on Information Systems 22(1), 5–53 (2004)

Capturing Buying Behaviour Using a Layered User Model

Oshadi Alahakoon[1], Seng Loke[2], and Arkady Zaslavsky[1]

[1] Caulfield School of Information Technology, Monash University, Australia
{oshadi,arkady.zaslavsky}@infotech.monash.edu.au
[2] Department of Computer Science and Computer Engineering, La Trobe University, Australia
s.loke@latrobe.edu.au

Abstract. User models are important tools for personalization, especially in ecommerce applications. But capturing dynamically changing user needs is a challenge. One of the reasons for such difficulty is that the purchasing behavior of an individual is based on a number of different aspects. In this paper we identify these aspects as a combination of demographics, domain based expectations and transactions. Since each individual can demonstrate a unique combination of these aspects, to achieve finer personalization, such individuality will have to be captured in user models. In this paper we propose such a user model architecture, which also has the ability to self-improve adapting to changes of individual behavior and long term modeling possibilities.

1 Introduction

Ecommerce websites can benefit greatly by offering personalized interactions. Such personalization is achieved by employing user models. Although the techniques used to build existing user models provided solutions for many requirements of user modeling, there are several issues that have yet to be satisfactorily addressed [3].One of the most important of such requirements is to identify 'reasons' for purchase decisions. As a solution, the work by Towle and Quinn [11] highlighted the importance of using knowledge based techniques to link user preferences to matching products. Knowledge based and utility based techniques have partially addressed these issues, but they lack another important requirement of user modeling ie. long term profiling capability [3].

Considering the above, we support the ideas in early work [3,12] regarding the advantages of explicit user-and product models and the use of the knowledge based approach to meaningfully link them. We argue that although purchasing behavior is based on a combination of demographics, domain based expectations and spontaneous transactions, each individual can demonstrate a unique combination of these aspects.

Section 2 of the paper describes different types of information representing users. Section 3 presents the architecture of the proposed user model while section 4 discusses details of its implementation. Section 5 describes the experiments which demonstrate the functionality of the model. Finally the conclusions are presented in section 6.

G. Psaila and R. Wagner (Eds.): EC-Web 2007, LNCS 4655, pp. 109–118, 2007.
© Springer-Verlag Berlin Heidelberg 2007

2 Identifying and Segmenting User Buying Behavior for Personalization

Modeling individual purchasing behavior is a highly complex task. Rather than attempting to model such complexity, we have worked on capturing the elements, which contribute to such behavior. Once captured, we focus on combining the elements together for individuals, creating more individualized profiles. We have identified three categories, which make up the aggregate purchasing behavior of individuals. (i). demography based general consumer buying behavior, (ii) buying preferences in each domain and (iii) transaction based needs in each interaction. These are discussed and justified below.

The VALS survey [13] (by the Stanford Research Institute) has defined eight consumer segments measuring individual lifestyle choices, and cultural and demographic aspects. Lifestyle Finder [9], too believe in static market segments. In many user modelling applications such static segments are used as stereotypes when more individualised information is unavailable [2,11]. Work carried out in [10] believes that static market segments are a result of a dynamic equilibrium, where consumers tend to change segments. Therefore the individual user expectations are needed to be identified and catered for rather than catering for different consumer segments. The above mentioned model also discusses the expectations of benefits (of products) of individuals. Among many market segmentation methods, benefit segmentation takes the customer's perspective and have proven to perform well [8]. By considering above work we combine demographics and benefit expectations to obtain an initial understanding of the users.

Although generally consumer behavior is expected to be in accordance with demographics, individuals have different opinions, and priorities about different product categories. These differences can result in deviations from the expected characteristic based behaviors as inferred by placing an individual within a certain stereotype. Identifying and modeling such individuality enhances personalization. Even when a user shows continuous behavioral patterns in a given domain, sudden mood changes or unusual events in their lives can cause deviations from such behaviors. For example an individual generally buying inexpensive clothing may compel by an advertisement to tryout something expensive. Therefore, keeping track of individual transactions will help not only to capture an individual's purchasing pattern in a given domain, but also his/her tendency towards unusual spontaneous purchasing.

Ones these three types of information are gathered in three separate layers, Individuals can be treated according to their dominating behavior. Buyers who behave according to the demographic based expectations will show a more consistent behavior across all domains. Some individuals will have varying expectations across product domains but continuity within a product domain. Third type of buyers will show frequently changing behavior and can be provided with current request based needs, rather than inferred facts from previous behavior.

3 Layered User Model Architecture

To implement the above ideas we propose a user model consisting of three layers, as shown in figure 1. Layer 1 of the user model contains (i) user identification details,

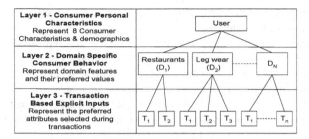

Fig. 1. Three layered User model architecture

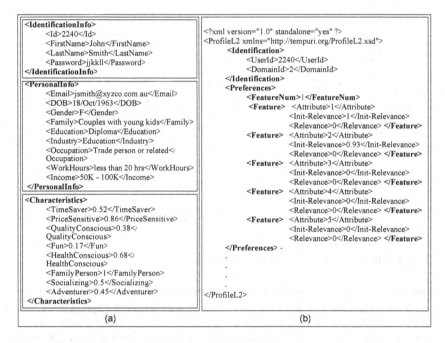

Fig. 2. (a) Layer 1 generated for user John Smith and (b) shows a section of the layer 2 for the same user with initially calculated relevance values

(ii) user's personal information obtained at the registration by filling a form, and (iii) buying characteristics calculated using demographics.

Layer 2 consists of user identification to facilitate linking between other layers, and derived relevance values indicating the importance of the attributes to the user.

Domain Features ($d_i f_j$ - j^{th} feature of domain i) are product descriptions that are important to consumers when deciding which product to purchase. A given domain d_i is made up of one or more *domain features* (eg. for a footwear domain, features are Cost, Size, etc). A domain feature consists of domain *attributes* (the cost feature has domain attributes corresponding to cost ranges: Low, Low-Medium, Medium, Med_High and High, where the unit price is <= 5, 5-10, 10-15, 15-25, and >25 respectively). Figure 2 shows an example of a layer 1 and a section of the layer 2 (in footwear domain) for the same user.

When the user transacts in a new domain, an individualized second layer for that domain is created. The layer 2 relevance values are initially derived from layer 1 user characteristics and later updated with actual transactions. When a user is interested in an item (or items) from a large collection of items, such selection involves asking a series of questions related to item attributes. Users' responses (consist of preferred attribute values) to such questions are collected in layer 3 as transaction information. Such layer 3 information is used to incrementally update the respective layer 2. The steps involved in populating and updating the user model layers is summarized below as a flow chart.

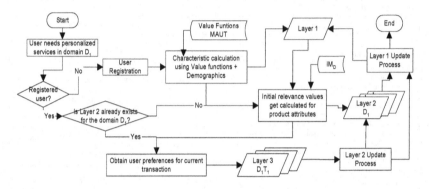

Fig. 3. The steps involved in creation and update of the layers

4   Implementation of the Architecture

Calculation of character values (Layer1). Each user characteristic is initially represented by a *value function* based on demographics. In our prototype system we use eight demographics (first column, Figure 4) combined within eight *value functions* (first row, Figure 4). We considered demographic types in other applications such as the Vals System [13] , LifeStyle Finder [9], and Australian Census data [4]. The *value functions* are defined using Multi Attribute Utility Theory (MAUT) combining demographics to generate an estimated value for each user characteristic.

For each user x ; $\exists$ $ch_i(x)$ where (i= 1, 2... n) and $\forall i$ $1 \geq ch_i(x) \geq 0$

Using MAUT, for an individual x , his/her i^{th} characteristic $ch_i(x)$ is given by:

$$ch_i(x) = \sum_{j=1}^{n} w_j v_j(x) , \text{ where } \sum_{j=1}^{n} w_j = 1 , \ v_j(x) \text{ is the evaluation of a given characteristic}$$

on the jth value dimension. In this case each demographic becomes a value dimension. Therefore $v_j(x)$, is a demographic value for demographic attribute j , and w_j is the weight determining the impact of the demographic value on the overall evaluation. In other words w_j is the relative importance of a demographic attribute in determining a given characteristic. n is the number of different demographic attributes related to the characteristic i, and $n \geq 0$. Demographic attributes are related to a characteristic based on common sense heuristics. For example *Time Saver* characteristic depends on an individual's time spent on 'work' and 'family'. Therefore the work related attributes (work hours, occupation) and family related attributes (family type, gender, and age group) are combined using different weights to calculate the *Time Saver* characteristic. In the calculations resulting characteristics are scaled between 0-1. The eight user characteristics and the contributing demographic attributes are shown in figure 4.

	Time saver	Price sensitive	Quality conscious	Fun spending	Health conscious	Family person	Socializing	Adventurer
Family Type	X	X		X		X	X	X
Gender	X				X			
Work Hours	X							
Age Group	X	X		X	X	X	X	X
Income		X	X	X			X	X
Occupation	X		X				X	
Industry				X			X	
Education					X			

Fig. 4. Demographics mapped to characteristics

Populating Layer 2 with the product attribute relevance values. Initial values for layer 2 are derived using an *'influence matrix'*. Each domain d_l (the lth domain) has an *influence matrix* (IM_{d_l}), where columns represent buying characteristics ch_i, and the weights w_i of the ith buying characteristic influencing a given attribute (at_k). Rows represent attributes of the particular domain. As such, value ranges in the IM, im_{at_k, ch_i} (where k is the attribute number and i is the characteristic number, and i,k =1,2,3,...n) represent the possible value ranges for characteristic $ch_i(x)$ of a user x, who is expected to show interest in attribute at_k. Since for each individual, behavioural characteristics are different, the relevance of an attribute varies from one user to another. The IM_{d_l} is now used to populate layer 2 of the user x. Characteristic values ($ch_i(x)$) for user x are mapped on to IM_{d_l} to produce relevance levels for domain attributes (at_k), and stored in layer 2. A section of the $IM_{footwear}$ (for the footwear domain) is shown in figure 5.

When calculating initial relevance values, two different methods were followed for continuous and discrete attributes. When the attributes are continuous, it is not fair to assign a crisp relevance value. Therefore the initial relevance values are calculated using fuzzy membership functions (figure 6(a)). This way more than one attribute belonging to the given feature becomes relevant to an individual depending on his/her characteristic value. As shown in figure 6(a), an individual with price sensitivity 0.86, is expected to prefer more than one cost ranges (Low, Low-Medium) in different degrees (1 and .93 respectively). By doing so, the assumption we make about the user becomes more flexible rather than strictly assigning a preference for the user.

Relevance of a discrete attribute is calculated depending on its *direction of influence* or in other words positive or negative influence. In figure 5, both *Quality Consciousness* (QC) and *Health Consciousness* (HC) characteristics are considered as positively influencing the preference for 'cotton' material. This implies individuals with high HC and QC values (where both values in 0.65-1 range) are likely to prefer cotton footwear.

Feature	Feature Name	Feature Type	Attribute number	Quality Conscious	QcWeight	Price Sensitivity	psWeight	Health Conscious	HcWeight	Socialising	soWeight
1	Cost	c	Low Price			.85,1		0	0	0	0
1	Cost	c	Low-Medium Price			.7,.85		0	0	0	0
1	Cost	c	Medium Price			.55,.7		0	0	0	0
1	Cost	c	MedHigh price			.35,.55		0	0	0	0
1	Cost	c	High Price			0,.35		0	0	0	0
10	Material	d	Cotton	0.65,1	.4	0		.65,1	.6	0	0
10	Material	d	Silk	0.65,1	.5	0		0		.65,1	.5

Fig. 5. A section of the Influence Matrix for footwear domain. For example, row 4 says that continuous attribute 'Med High Price' belonging to feature number 1 (cost) becomes relevant to an individual if his/her PS value is in the .35 to .55 range.

Once the layer 2 is available and populated with initial relevance values for domain attributes, the user can specify an initial query. The querying process and how the product search is done are out of scope of this paper. At the end of the search, the user's explicit preference values obtained during the interactive search process is used in updating the layer 2 of the user model. The explicit preferences are recorded in the layer 3 of the user model as attribute and relevance value pairs.

Updating attribute relevance values with Layer 3 transaction information. We use Hebbian learning to capture these individual transactions and update the respective attribute relevance values in layer 2. According to Hebb [5], "When a neuron A repeatedly and persistently take part in exciting neuron B , the synaptic connection between A and B will be strengthened". So a Hebbian network can be used as an associator which will establish the association between two sets of patterns $\{X_i, i = 1,....,L \}$ and $\{Y_j, j = 1,.., L\}$. We use this technique to update layer 2 relevance values according to the changes in attribute values from layer 3. Since each feature is connected to one or more attributes, our output layer has only one node. Therefore, the new relevance rel_{new} is given by $rel_{new} = rel_{old} + \eta(i - rel_{old})$ where rel_{old} is the existing relevance, i is the new input value (binary 0/1) and $0 > \eta > 1$ is the learning rate. This method gradually changes an individual's preference towards a given attribute either by decrementing or incrementing the relevance value of the attribute. For example user John Smith is expected to prefer the 'Low' price range according to his characteristic based stereotypic behaviour. If he continue to purchase the goods in the same range then the

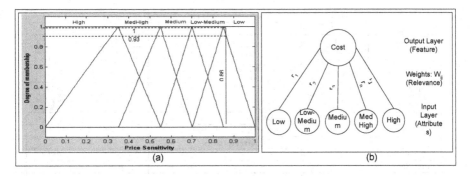

Fig. 6. (a) Continuous feature 'Cost' has five membership functions for its five attributes. For example if the PS characteristic of the user is 0.86, then two of the attributes (Low-1, LowMedium -0.93) get initial relevance values. (b) For example 'Cost' feature is connected to five attributes. Attribute with the highest relevance becomes the most suitable value for the Feature.

relevance for that attribute grows (the connection with the 'Low' node get strengthen). The relevance values for other attributes of the feature get weakened.

If he purchase an item in the highest price range due to a special reason, such occurrence do not mean hereafter he is going to stick to that new price range. The advantage of this method is such sudden behaviour changes will have less impact on the general user trend, since the relevance is changed only from a η amount where $0 > \eta > 1$.

5 Demonstration of the Model

For our experimentation and evaluation purposes, we use a dataset about footwear obtained from KDD-CUP 2000 [7]. This data consists of actual purchasing transactions for 3466 customers and their demographics. A subset of this data is used to demonstrate our model. Initially demographic values were used to generate the individual characteristics as described in section 4.

Figure 7 shows the outcome for five selected individuals. Users 1, 2, 3, and 4 are demographically different from each other, while user 4 is very similar to user 5. User 1 having a young family and middle income is considered 'price sensitive' and have a low value for 'fun spending'. Being a bachelor, user 2 is less price sensitive than user 1, although with low income. User 3 without kids and high income has contrasting values for the same characteristics. Users 4 & 5 being demographically similar, have the same characteristic values.

Due to demographic similarity and as such derived characteristic similarity, the initial relevance values for user 4 and 5 are the same. But after the purchase transactions (27 interactions for user 4 and 18 interactions for user 5) their relevance values for several of the attributes show a significant change. Predicted relevance values for cost (1-5 in 8(a)) and preference for fashion/basic leg wear (25-27) remains closer to the original predicted values. This supports the accuracy of demographic based predictions. Attributes 6-24 represent the popularity levels of footwear brands. It is shown that both users behaved different from the predictions. This shows dissimilarity in

	User 1	User 2	User 3	User 4	User 5
Age	37	25	55	40	38
Gender	F	M	M	F	F
Family	Couples with young kids	Single/Bachelor	Couples without kids	Couples with young kids	Couples with young kids
Education	Batchelor Degree	Diploma/Advanced Diploma	Diploma/Advanced Dip	High School	High School
Industry	Computers	Construction	Retail	Education	Automotive
Occupation	Associate Professional	Trade person or related	Managerial/Administration	Trade person or related	Trade person or related
Work Hours	21-40 hrs	21-40 hrs	40 hrs & more	20 hrs & less	20 hrs & less
Income	50K - 100K	30K - 50K	100K - 200K	50K - 100K	50K - 100K
Characteristics					
Time Saver	0.71	0.42	0.65	0.52	0.52
Price Sensitive	0.85	0.53	0.3	0.86	0.86
Q/Conscious	0.52	0.29	0.81	0.38	0.38
Fun Spending	0.03	0.61	0.71	0.17	0.17
H/Conscious	0.72	0.19	0.84	0.68	0.68
Family Person	1	0.05	0.34	1	1
Socializing	0.4	0.3	0.47	0.5	0.5
Adventurer	0.45	0.91	0.54	0.45	0.45

Fig. 7. Demographics and characteristic values for five individuals

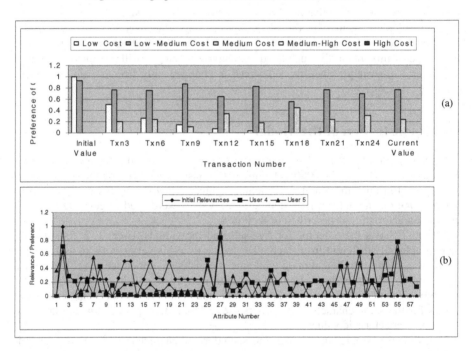

Fig. 8. (a) Initial and after transactions (changed) relevance values for two demographically similar users (4 & 5) (b) Change of relevance values for cost ranges for User 4, during the 28 transactions

purchasing of demographically similar users. These observations demonstrate the ability of the model to capture both demographic based similarities and dissimilarities in purchasing behaviour. Attributes which are difficult to predict such as "size"

(25-47) has initial zero values and gain relevance values with transactions. This shows the ability of the model in capturing undefined user preferences which are difficult to predict based on demographics.

Figure 8(b) shows how initial relevance for lowest price range 'low cost' becomes zero and how a new price range 'medium cost' becomes relevant for user 4. This highlights a change in the demographic based prediction and the ability of the model to adapt to changing behaviour. Experiments demonstrate the functionality of the model. Initially calculated relevance values are confirmed, and some are contradicted. Several attributes (eg. size, pattern) which did not have initial relevance values received new values from actual transactions.

6 Conclusions

This new user model architecture has been designed to be implemented within the prototype personalisation component of a multi-agent system called eHermes [6,1]. The model has the ability to capture three types of information (i) demographic based general consumer buying behavior, (ii) long term buying preferences in each domain, and (iii) transaction based needs for each interaction. Since the layered user model has the ability to describe users from three different aspects, it has the flexibility to understand the unique behavior of different individuals. With long-term usage it can classify a user as a buyer behaving according to demographics, a buyer prefers to follow fixed purchased trends within a given domain or as a buyer who makes more spontaneous buying decisions. Such information will be valuable in recommendation and marketing tasks.

Among existing user models, we identify the user models in [2] for personalized EPG for TV and Personis [7] user modeling server as having based on similar ideas. 'Personis' [7] user modelling server maintains a main user model and a collection of application based 'Personas' for each individual. Since these personas hold application domain specific information, they seem conceptually similar to the second layer of information in our layered user model. The main difference with Personis user models is our proposed architecture has the ability to infer domain dependent behaviour information from layer 1 according to the system knowledge where in Personis such information is based on self assessments of their abilities from the users. Personis does not use stereotypes. Although the user model used in [2] is not reusable for other domains, the entire architecture is organized as a 'shell', which is easy to adapt for a new domain. But each new domain needs knowledge reengineering such as building a new set of stereotypes. In our architecture, initial second layer values are obtained using a similar strategy as stereotypes. Registration information is used to calculate behavioural characteristics and those values are used against an influence matrix to obtain initial domain model of the user. Since the characteristics values are reusable irrespective of the domain, only the influence matrix requires recreation for each different domain. This needs less effort than generating an entire set of stereotypes for each new domain. EPGs in [2] map the user's demographics to predefined stereotypes (eg. Housewife) and directly assign initial user preferences. Stereotyping in the proposed model is more flexible as the characteristics and the influence matrix are separately used. This way, the influence matrix could be updated by analysing long term behaviours of the users. But the stereotypes described in [2] are based on outcome of the survey and therefore will remain fixed.

In our experimentation we have analyzed the constancy, spontaneity of individuals, and domain based purchase peculiarities for users. Such information is valuable to use as domain rules for updating IMs' or to update value functions used in personal characteristic generation. Such possibilities are currently being investigated.

References

1. Alahakoon, O., Loke, S., Zaslavsky, A.: Use of Buying Behavioural Characteristics in Long Term User Models. In: Proceedings of the 2nd International Workshop on Web Personalisation, Recommender Systems and Intelligent User Interfaces (WPRSIUI 2006) at the 4th International Conference on Adaptive Hypermedia (AH 2006). Dublin, Ireland (June 2006)
2. Ardissono, L., Gena, C., Torasso, P., Bellifemine, F., Difino, A., Negro, B.:User Modelling and Recommendation Techniques for Personalized Electronic Program Guides. In Personalized Digital Television: Targeting programs to individual users. In: Ardissono, L., Kobsa, A., Maybury, M. (eds.), pp. 71–101. Kluwer Academic, Dordrecht (2004)
3. Burke, R.: Hybrid Reccommender Systems: Survey and Experiments. User Modelling and User-Adapted Interaction 12, 331–370 (2002)
4. Census_Data.: A Snapshot of Victoria, Census Basic Community Profile and Snapshot. Australian Bureau of Statistics (2001)
5. Hebb, D.: The Organization of Behavior. Wiley, NewYork (1949)
6. Jayaputera, G.T., Zaslavsky, A., Loke, S.W.: In: IEEE/WIC/ACM International Conference on Intelligent Agent Technology (IAT2004), Beijing, China, pp. 79–85. IEEE Computer Society, Los Alamitos (2004)
7. Kay, J., Kummerfeld, R.J., Lauder, P.: In Adaptive Hypermedia. In: De Bra, P., Brusilovsky, P., Conejo, R. (eds.) AH 2002. LNCS, vol. 2347, pp. 203–212. Springer, Heidelberg (2002)
8. Kelly, S.: Customer Intelligence from Data to Dialogue. John Wiley & Sons Ltd, West Sussex, England (2006)
9. Krulwich, B.: Lifestyle finder: intelligent user profiling using large-scale demographic data. AI Magazine 18(2), 37–45 (1997)
10. Patel, S., Schlijper, A.: Models of Consumer behaviour, Unilever (2003) http://www.maths.ox.ac.uk/ociam/StudyGroups/ESGI49/problems/unilever2/unilever2.pdf
11. Rich, E.: User modeling via stereotypes. Cognitive Science 3, 329–354 (1979)
12. Towle, B., Quinn, C.: In: Knowledge-Based Electronic Markets, Papers from the AAAI Workshop, AAAI Technical Report WS-00-04AAAI Press, Menlo Park, CA, pp. 74–77 (2000)
13. VALS-system, SRI Consulting, The Valve and LifeStyle survey www.srlc-bi.com/VALS/presurvey.shtml

Building Business Relationships with Negotiation

John Debenham[1] and Carles Sierra[2]

[1] University of Technology, Sydney, Australia
debenham@it.uts.edu.au
[2] Institut d'Investigacio en Intel.ligencia Artificial, Spanish Scientific Research Council, UAB
08193 Bellaterra, Catalonia, Spain
sierra@iiia.csic.es

Abstract. Successful negotiators prepare by determining their position along five dimensions. We introduce a negotiation model based on these dimensions and on two primitive concepts: *intimacy* (degree of closeness) and *balance* (degree of fairness). The *intimacy* is a pair of matrices that evaluate both an agent's contribution to the relationship and its opponent's contribution each from an information view and from a utilitarian view. The *balance* is the difference between these matrices. A *relationship strategy* maintains a *target intimacy* for each relationship that an agent would like the relationship to move towards in future. The *negotiation strategy* maintains a set of Options that are in-line with the current intimacy level, and then *tactics* wrap the Options in argumentation with the aim of attaining a successful deal *and* manipulating the successive negotiation balances towards the target intimacy.

1 Introduction

In this paper we propose a new negotiation model to deal with long term relationships that are founded on successive negotiation encounters. The model is grounded on results from business and psychological studies [1,2,3], and acknowledges that negotiation is an information exchange process as well as a utility exchange process [4,5]. We believe that if agents are to succeed in real application domains they have to reconcile both views: informational and game-theoretical. Before a negotiation starts human negotiators prepare the dialogic exchanges that can be made along the five LOGIC dimensions [6]:

- Legitimacy. What information is relevant to the negotiation process? What are the persuasive arguments about the fairness of the options?
- Options. What are the possible agreements we can accept?
- Goals. What are the underlying things we need or care about? What are our goals?
- Independence. What will we do if the negotiation fails? What alternatives have we got?
- Commitment. What outstanding commitments do we have?

Our aim is to model trading scenarios where agents represent their human principals, and thus we want their behaviour to be comprehensible by humans and to respect usual human negotiation procedures, whilst being consistent with, and somehow extending,

G. Psaila and R. Wagner (Eds.): EC-Web 2007, LNCS 4655, pp. 119–128, 2007.
© Springer-Verlag Berlin Heidelberg 2007

game theoretical and information theoretical results. In this sense, agents are not just utility maximisers, but aim at building long lasting relationships with progressing levels of *intimacy* that determine what *balance* in information and resource sharing is acceptable to them. These two concepts, intimacy and balance are key in the model, and enable us to understand competitive and co-operative game theory as two particular theories of agent relationships (i.e. at different intimacy levels). These two theories are too specific and distinct to describe how a (business) relationship might grow because interactions have some aspects of these two extremes on a continuum in which, for example, agents reveal increasing amounts of private information as their intimacy grows. Negotiation strategies can then naturally be seen as procedures that select tactics used to attain a successful deal *and* to reach a *target* intimacy level. It is common in human settings to use tactics that compensate for unbalances in one dimension of a negotiation with unbalances in another dimension. In this sense, humans aim at a *general sense of fairness* in an interaction.

In Section 2 we outline the aspects of human negotiation modelling that we cover in this work. Section 3 explains in outline the architecture and the concepts of intimacy and balance, and how they influence the negotiation. Section 4 contains a description of the different metrics used in the agent model including intimacy. Finally, Section 5 outlines how strategies and tactics use the LOGIC framework, intimacy and balance.

2   Human Negotiation

Negotiation dialogues exchange dialogical moves, i.e. messages, with the intention of getting information about the opponent or giving away information about us along these five dimensions: request for information, propose options, inform about interests, issue promises, appeal to standards ... A key part of any negotiation process is to build a model of our opponent(s) along these dimensions. All utterances agents make during a negotiation give away information about their current LOGIC model, that is, about their legitimacy, options, goals, independence, and commitments. Also, several utterances can have a utilitarian interpretation in the sense that an agent can associate a preferential gain to them. For instance, an offer may inform our negotiation opponent about our willingness to sign a contract in the terms expressed in the offer, and at the same time the opponent can compute what is its associated expected utilitarian gain. These two views: information-based and utility-based, are central in the model proposed in this paper.

2.1   Intimacy and Balance in Relationships

There is evidence from psychological studies that humans seek a *balance* in their negotiation relationships. The classical view [1] is that people perceive resource allocations as being distributively fair (i.e. well balanced) if they are proportional to inputs or contributions (i.e. equitable). However, more recent studies [2,7] show that humans follow a richer set of norms of distributive justice depending on their *intimacy* level: equity, equality, and need. *Equity* being the allocation proportional to the effort, *equality* being the allocation in equal amounts, and *need* being the allocation proportional to the need for the resource. For instance, if we are in a purely economic setting (low intimacy) we might request equity for the Options dimension but could accept equality in the Goals dimension.

The perception of a relation being in balance (i.e. fair) depends strongly on the nature of the social relationships between individuals (i.e. the intimacy level). In purely economical relationships (e.g., business), equity is perceived as more fair; in relations where joint action or fostering of social relationships are the goal (e.g. friends), equality is perceived as more fair; and in situations where personal development or personal welfare are the goal (e.g. family), allocations are usually based on need.

We believe that the perception of balance in dialogues (in negotiation or otherwise) is grounded on social relationships, and that every dimension of an interaction between humans can be correlated to the social closeness, or *intimacy*, between the parties involved. According to the previous studies, the more intimacy across the five LOGIC dimensions the more the need norm is used, and the less intimacy the more the equity norm is used. This might be part of our social evolution. There is ample evidence that when human societies evolved from a hunter-gatherer structure[1] to a shelter-based one[2] the probability of survival increased when food was scarce.

The perceived balance in a negotiation dialogue allows negotiators to infer information about their opponent, about its LOGIC stance, and to compare their relationships with all negotiators. For instance, if we perceive that every time we request information it is provided, and that no significant questions are returned, or no complaints about not receiving information are given, then that probably means that our opponent perceives our social relationship to be very close. Alternatively, we can detect what issues are causing a burden to our opponent by observing an imbalance in the information or utilitarian senses on that issue.

3 Agent Architecture

A multiagent system $\{\alpha, \beta_1, \ldots, \beta_n, \xi, \theta_1, \ldots, \theta_t\}$, contains an agent α that interacts with other *argumentation agents*, β_i, *information providing agents*, θ_j, and an *institutional agent*, ξ, that represents the institution where we assume the interactions happen [8]. The institutional agent reports promptly and honestly on what actually occurs after an agent signs a contract, or makes some other form of commitment. In Section ?? this enables us to measure the difference between an utterance and a subsequent observation. Agents have a probabilistic first-order *internal language* L used to represent a *world model*, $\mathcal{M}^t$. A generic *information-based* architecture is described in detail in [4].

Agent α acts in response to a *need* that is expressed in terms of the ontology. A need may be exogenous such as a need to trade profitably and may be triggered by another agent offering to trade, or endogenous such as α deciding that it owns more wine than it requires. Needs trigger α's goal/plan proactive reasoning, while other messages are

---

[1] In its purest form, individuals in these societies collect food and consume it when and where it is found. This is a pure equity sharing of the resources, the gain is proportional to the effort.

[2] In these societies there are family units, around a shelter, that represent the basic food sharing structure. Usually, food is accumulated at the shelter for future use. Then the food intake depends more on the need of the members.

dealt with by α's reactive reasoning.[3] Each plan prepares for the negotiation by assembling the contents of a 'LOGIC briefcase' that the agent 'carries' into the negotiation[4]. The *relationship strategy* determines which agent to negotiate with for a given need; it uses risk management analysis to preserve a strategic set of trading relationships for each mission-critical need — this is not detailed here. For each trading relationship this strategy generates a *relationship target* that is expressed in the LOGIC framework as a desired level of *intimacy* to be achieved in the long term.

Each negotiation consists of a dialogue, Ψ^t, between two agents with agent α contributing utterance μ and the partner β contributing μ'. Each dialogue, Ψ^t, is evaluated using the LOGIC framework in terms of the *value* of Ψ^t to both α and β — see Section 4.2. The *negotiation strategy* then determines the current set of Options $\{\delta_i\}$, and then the *tactics*, guided by the *negotiation target*, decide which, if any, of these Options to put forward and wraps them in argumentation dialogue — see Section 5. We now describe two of the distributions in $\mathcal{M}^t$ that support offer exchange.

$\mathbb{P}^t(\mathrm{acc}(\alpha,\beta,\chi,\delta))$ estimates the probability that α should accept proposal δ in satisfaction of her need χ, where $\delta = (a,b)$ is a pair of commitments, a for α and b for β. α will accept δ if: $\mathbb{P}^t(\mathrm{acc}(\alpha,\beta,\chi,\delta)) > c$, for level of certainty c. This estimate is compounded from subjective and objective views of acceptability. The *subjective estimate* takes account of: the extent to which the enactment of δ will satisfy α's need χ, how much δ is 'worth' to α, and the extent to which α believes that she will be in a position to execute her commitment a[5,4]. $S_\alpha(\beta,a)$ is a random variable denoting α's estimate of β's subjective valuation of a over some finite, numerical evaluation space. The *objective estimate* captures whether δ is acceptable on the open market, and variable $U_\alpha(b)$ denotes α's open-market valuation of the enactment of commitment b, again taken over some finite numerical valuation space. We also consider needs, the variable $T_\alpha(\beta,a)$ denotes α's estimate of the *strength* of β's motivating *need* for the enactment of commitment a over a valuation space. Then for $\delta = (a,b)$: $\mathbb{P}^t(\mathrm{acc}(\alpha,\beta,\chi,\delta)) =$

$$\mathbb{P}^t\left(\left(\frac{T_\alpha(\beta,a)}{T_\alpha(\alpha,b)}\right)^h \times \left(\frac{S_\alpha(\alpha,b)}{S_\alpha(\beta,a)}\right)^g \times \frac{U_\alpha(b)}{U_\alpha(a)} \geq s\right) \tag{1}$$

where $g \in [0,1]$ is α's *greed*, $h \in [0,1]$ is α's degree of *altruism*, and $s \approx 1$ is derived from the *stance*[5] described in Section 5. The parameters g and h are independent. We can imagine a relationship that begins with $g = 1$ and $h = 0$. Then as the agents share increasing amounts of their information about their open market valuations g gradually reduces to 0, and then as they share increasing amounts of information about their needs

---

[3] Each of α's plans and reactions contain constructors for an initial *world model* $\mathcal{M}^t$. $\mathcal{M}^t$ is then maintained from percepts received using *update functions* that transform percepts into constraints on $\mathcal{M}^t$ — for details, see [5,4].

[4] Empirical evidence shows that in human negotiation, better outcomes are achieved by skewing the opening Options in favour of the proposer. We are unaware of any empirical investigation of this hypothesis for autonomous agents in real trading scenarios.

[5] If α chooses to inflate her opening Options then this is achieved in Section 5 by increasing the value of s. If $s \gg 1$ then a deal may not be possible. This illustrates the well-known inefficiency of bilateral bargaining established analytically by Myerson and Satterthwaite in 1983.

h increases to 1. The basis for the acceptance criterion has thus developed from equity to equality, and then to need.

$\mathbb{P}^t(\text{acc}(\beta,\alpha,\delta))$ estimates the probability that β would accept δ, by observing β's responses. For example, if β sends the message $\text{Offer}(\delta_1)$ then α derives the constraint: $\{\mathbb{P}^t(\text{acc}(\beta,\alpha,\delta_1)) = 1\}$ on the distribution $\mathbb{P}^t(\beta,\alpha,\delta)$, and if this is a counter offer to a former offer of α's, δ_0, then: $\{\mathbb{P}^t(\text{acc}(\beta,\alpha,\delta_0)) = 0\}$. In the not-atypical special case of multi-issue bargaining where the agents' preferences over the individual issues *only* are known and are complementary to each other's, maximum entropy reasoning can be applied to estimate the probability that *any* multi-issue δ will be acceptable to β by enumerating the possible worlds that represent β's "limit of acceptability" [9].

4   Summary Measures

A *dialogue*, Ψ^t, between agents α and β is a sequence of inter-related utterances in context. A *relationship*, Ψ^{*t}, is a sequence of dialogues. We first measure the *confidence* that an agent has for another by observing, for each utterance, the difference between what is said (the utterance) and what subsequently occurs (the observation). Second we *evaluate* each dialogue as it progresses in terms of the LOGIC framework — this evaluation employs the confidence measures. Finally we define the *intimacy* of a relationship as an aggregation of the value of its component dialogues.

4.1   Confidence

Confidence measures generalise what are commonly called *trust*, *reliability* and *reputation* measures into a single computational framework that spans the LOGIC categories. In Section 4.2 confidence measures are applied to valuing fulfilment of promises in the Legitimacy category — we formerly called this "honour" [5], to the execution of commitments — we formerly called this "trust" [10], and to valuing dialogues in the Goals category — we formerly called this "reliability" [5].

Ideal observations. Consider a distribution of observations that represent α's "ideal" in the sense that it is the best that α could reasonably expect to observe. This distribution will be a function of α's *context* with β denoted by e, and is $\mathbb{P}^t_I(\varphi'|\varphi, e)$. Here we measure the relative entropy between this ideal distribution, $\mathbb{P}^t_I(\varphi'|\varphi, e)$, and the distribution of expected observations, $\mathbb{P}^t(\varphi'|\varphi)$. That is:

$$\mathbb{C}(\alpha, \beta, \varphi) = 1 - \sum_{\varphi'} \mathbb{P}^t_I(\varphi'|\varphi, e) \log \frac{\mathbb{P}^t_I(\varphi'|\varphi, e)}{\mathbb{P}^t(\varphi'|\varphi)} \tag{2}$$

where the "1" is an arbitrarily chosen constant being the maximum value that this measure may have. This equation measures confidence for a single statement φ. It makes sense to aggregate these values over a class of statements, say over those φ that are in the ontological context o, that is $\varphi \leq o$:

$$\mathbb{C}(\alpha, \beta, o) = 1 - \frac{\sum_{\varphi:\varphi\leq o} \mathbb{P}^t_\beta(\varphi)\,[1 - \mathbb{C}(\alpha, \beta, \varphi)]}{\sum_{\varphi:\varphi\leq o} \mathbb{P}^t_\beta(\varphi)}$$

where $\mathbb{P}_\beta^t(\varphi)$ is a probability distribution over the space of statements that the next statement β will make to α is φ. Similarly, for an overall estimate of β's *confidence* in α:

$$\mathbb{C}(\alpha,\beta) = 1 - \sum_\varphi \mathbb{P}_\beta^t(\varphi)\left[1 - \mathbb{C}(\alpha,\beta,\varphi)\right]$$

Preferred observations. The previous measure requires, $\mathbb{P}_I^t(\varphi'|\varphi,e)$, an ideal distribution for each φ. Here we measure the extent to which the observation φ' is preferable to the original statement φ. Given a predicate $\text{Prefer}(c_1,c_2,e)$ meaning that α prefers c_1 to c_2 in environment e. Then if $\varphi \leq o$:

$$\mathbb{C}(\alpha,\beta,\varphi) = \sum_{\varphi'} \mathbb{P}^t\left(\text{Prefer}(\varphi',\varphi,o)\right)\mathbb{P}^t(\varphi'|\varphi)$$

and:

$$\mathbb{C}(\alpha,\beta,o) = \frac{\sum_{\varphi:\varphi\leq o}\mathbb{P}_\beta^t(\varphi)\mathbb{C}(\alpha,\beta,\varphi)}{\sum_{\varphi:\varphi\leq o}\mathbb{P}_\beta^t(\varphi)}$$

Certainty in observation. Here we measure the consistency in expected acceptable observations, or "the lack of expected uncertainty in those possible observations that are better than the original statement". Let: $\Phi_+(\varphi,o,\kappa) = \{\varphi' \mid \mathbb{P}^t(\text{Prefer}(\varphi',\varphi,o)) > \kappa\}$ where $\varphi \leq o$ for some constant κ, and:

$$\mathbb{C}(\alpha,\beta,\varphi) = 1 + \frac{1}{B^*}\cdot\sum_{\varphi'\in\Phi_+(\varphi,o,\kappa)}\mathbb{P}_+^t(\varphi'|\varphi)\log\mathbb{P}_+^t(\varphi'|\varphi)$$

where $\mathbb{P}_+^t(\varphi'|\varphi)$ is the normalisation of $\mathbb{P}^t(\varphi'|\varphi)$ for $\varphi' \in \Phi_+(\varphi,o,\kappa)$,

$$B^* = \begin{cases} 1 & \text{if } |\Phi_+(\varphi,o,\kappa)| = 1 \\ \log|\Phi_+(\varphi,o,\kappa)| & \text{otherwise} \end{cases}$$

As above we aggregate this measure for observations in a particular context o, and measure confidence as before.

Computational Note. The various measures given above involve extensive calculations. For example, Eqn. 2 contains $\sum_{\varphi'}$ that sums over *all* possible observations φ'. We obtain a more computationally friendly measure by appealing to the structure of the ontology, and the right-hand side of Eqn. 2 may be approximated to:

$$1 - \sum_{\varphi':\text{Sim}(\varphi',\varphi)\geq\eta}\mathbb{P}_{\eta,I}^t(\varphi'|\varphi,e)\log\frac{\mathbb{P}_{\eta,I}^t(\varphi'|\varphi,e)}{\mathbb{P}_\eta^t(\varphi'|\varphi)}$$

where $\mathbb{P}_{\eta,I}^t(\varphi'|\varphi,e)$ is the normalisation of $\mathbb{P}_I^t(\varphi'|\varphi,e)$ for $\text{Sim}(\varphi',\varphi) \geq \eta$, and similarly for $\mathbb{P}_\eta^t(\varphi'|\varphi)$. The extent of this calculation is controlled by the parameter η. An even tighter restriction may be obtained with: $\text{Sim}(\varphi',\varphi) \geq \eta$ *and* $\varphi' \leq \psi$ for some ψ.

4.2   Valuing Negotiation Dialogues

Suppose that a negotiation commences at time s, and by time t a string of utterances, $\Phi^t = \langle \mu_1, \ldots, \mu_n \rangle$ has been exchanged between agent α and agent β. This negotiation dialogue is evaluated by α in the context of α's world model at time s, $\mathcal{M}^s$, and the environment e that includes utterances that may have been received from other agents in the system including the information sources $\{\theta_i\}$. Let $\Psi^t = (\Phi^t, \mathcal{M}^s, e)$, then α estimates the *value* of this dialogue to itself in the context of $\mathcal{M}^s$ and e as a 2×5 array $V_\alpha(\Psi^t)$ where:

$$V_x(\Psi^t) = \begin{pmatrix} I_x^L(\Psi^t) & I_x^O(\Psi^t) & I_x^G(\Psi^t) & I_x^I(\Psi^t) & I_x^C(\Psi^t) \\ U_x^L(\Psi^t) & U_x^O(\Psi^t) & U_x^G(\Psi^t) & U_x^I(\Psi^t) & U_x^C(\Psi^t) \end{pmatrix}$$

where the $I(\cdot)$ and $U(\cdot)$ functions are information-based and utility-based measures respectively as we now describe. α estimates the *value* of this dialogue to β as $V_\beta(\Psi^t)$ by assuming that β's reasoning apparatus mirrors its own.

In general terms, the information-based valuations measure the reduction in uncertainty, or information gain, that the dialogue gives to each agent, they are expressed in terms of decrease in entropy that can always be calculated. The utility-based valuations measure utility gain are expressed in terms of "some suitable" utility evaluation function $\mathbb{U}(\cdot)$ that can be difficult to define. This is one reason why the utilitarian approach has no natural extension to the management of argumentation that is achieved here by our information-based approach. For example, if α receives the utterance "Today is Tuesday" then this may be translated into a constraint on a single distribution, and the resulting decrease in entropy is the information gain. Attaching a utilitarian measure to this utterance may not be so simple.

We use the term "2×5 array" loosely to describe V_α in that the elements of the array are lists of measures that will be determined by the agent's requirements. Table 1 shows a sample measure for each of the ten categories, in it the dialogue commences at time s and terminates at time t. In that Table, $\mathbb{U}(\cdot)$ is a suitable utility evaluation function, needs(β, χ) means "agent β needs the need χ", cho(β, χ, γ) means "agent β satisfies need χ by choosing to negotiate with agent γ", N is the set of needs chosen from the ontology at some suitable level of abstraction, T^t is the set of offers on the table at time t, com(β, γ, b) means "agent β has an outstanding commitment with agent γ to execute the commitment b" where b is defined in the ontology at some suitable level of abstraction, B is the number of such commitments, and there are $n + 1$ agents in the system.

4.3   Intimacy and Balance

The *balance* in a negotiation dialogue, Ψ^t, is defined as: $B_{\alpha\beta}(\Psi^t) = V_\alpha(\Psi^t) \ominus V_\beta(\Psi^t)$ for an element-by-element difference operator $\ominus$ that respects the structure of $V(\Psi^t)$. The *intimacy* between agents α and β, $I_{\alpha\beta}^{*t}$, is the pattern of the two 2×5 arrays V_α^{*t} and V_β^{*t} that are computed by an update function as each negotiation round terminates, $I_{\alpha\beta}^{*t} = \left(V_\alpha^{*t}, V_\beta^{*t} \right)$. If Ψ^t terminates at time t:

$$V_x^{*t+1} = \nu \times V_x(\Psi^t) + (1 - \nu) \times V_x^{*t} \tag{3}$$

Table 1. Sample measures for each category in $V_\alpha(\Psi^t)$. (Similarly for $V_\beta(\Psi^t)$.).

$$I_\alpha^L(\Psi^t) = \sum_{\varphi \in \Psi^t} \mathbb{C}^t(\alpha, \beta, \varphi) - \mathbb{C}^s(\alpha, \beta, \varphi)$$

$$U_\alpha^L(\Psi^t) = \sum_{\varphi \in \Psi^t} \sum_{\varphi'} \mathbb{P}_\beta^t(\varphi' | \varphi) \times \mathbb{U}_\alpha(\varphi')$$

$$I_\alpha^O(\Psi^t) = \frac{\sum_{\delta \in T^t} \mathbb{H}^s(\text{acc}(\beta, \alpha, \delta)) - \sum_{\delta \in T^t} \mathbb{H}^t(\text{acc}(\beta, \alpha, \delta))}{|T^t|}$$

$$U_\alpha^O(\Psi^t) = \sum_{\delta \in T^t} \mathbb{P}^t(\text{acc}(\beta, \alpha, \delta)) \times \sum_{\delta'} \mathbb{P}^t(\delta' | \delta) \mathbb{U}_\alpha(\delta')$$

$$I_\alpha^G(\Psi^t) = \frac{\sum_{\chi \in N} \mathbb{H}^s(\text{needs}(\beta, \chi)) - \mathbb{H}^t(\text{needs}(\beta, \chi))}{|N|}$$

$$U_\alpha^G(\Psi^t) = \sum_{\chi \in N} \mathbb{P}^t(\text{needs}(\beta, \chi)) \times \mathbb{E}^t(\mathbb{U}_\alpha(\text{needs}(\beta, \chi)))$$

$$I_\alpha^I(\Psi^t) = \frac{\sum_{i=1}^o \sum_{\chi \in N} \mathbb{H}^s(\text{cho}(\beta, \chi, \beta_i)) - \mathbb{H}^t(\text{cho}(\beta, \chi, \beta_i))}{n \times |N|}$$

$$U_\alpha^I(\Psi^t) = \sum_{i=1}^o \sum_{\chi \in N} \mathbb{U}^t(\text{cho}(\beta, \chi, \beta_i)) - \mathbb{U}^s(\text{cho}(\beta, \chi, \beta_i))$$

$$I_\alpha^C(\Psi^t) = \frac{\sum_{i=1}^o \sum_{\delta \in B} \mathbb{H}^s(\text{com}(\beta, \beta_i, b)) - \mathbb{H}^t(\text{com}(\beta, \beta_i, b))}{n \times |B|}$$

$$U_\alpha^C(\Psi^t) = \sum_{i=1}^o \sum_{\delta \in B} \mathbb{U}^t(\text{com}(\beta, \beta_i, b)) - \mathbb{U}^s(\text{com}(\beta, \beta_i, b))$$

where ν is the learning rate, and $x = \alpha, \beta$. Additionally, V_x^{*t} continually decays by: $V_x^{*t+1} = \tau \times V_x^{*t} + (1 - \tau) \times D_x$, where $x = \alpha, \beta$; τ is the decay rate, and D_x is a 2×5 array being the decay limit distribution for the value to agent x of the intimacy of the relationship in the absence of any interaction. D_x is the *reputation* of agent x. The *relationship balance* between agents α and β is: $B_{\alpha\beta}^{*t} = V_\alpha^{*t} \ominus V_\beta^{*t}$. In particular, the intimacy determines values for the parameters g and h in Equation 1. As a simple example, if both $I_\alpha^O(\Psi^{*t})$ and $I_\beta^O(\Psi^{*t})$ increase then g decreases, and as the remaining eight information-based LOGIC components increase, h increases.

The notion of balance may be applied to pairs of utterances by treating them as degenerate dialogues. In simple multi-issue bargaining the *equitable information revelation* strategy generalises the tit-for-tat strategy in single-issue bargaining, and extends to a tit-for-tat argumentation strategy by applying the same principle across the LOGIC framework.

5   Strategies and Tactics

Each negotiation has to achieve two goals. First it may be intended to achieve some contractual outcome. Second it will aim to contribute to the growth, or decline, of the relationship intimacy.

We now describe in greater detail the way in which these agents go about building relationships with negotiation. The negotiation literature consistently advises that an agent's behaviour should not be predictable even in close, intimate relationships. The required variation of behaviour is normally described as varying the negotiation *stance* that informally varies from "friendly guy" to "tough guy". The stance injects bounded random noise into the process, where the bound tightens as intimacy increases. The stance, $S_{\alpha\beta}^t$, is a 2×5 matrix of randomly chosen multipliers, each ≈ 1, that perturbs α's actions. The value in the (x,y) position in the matrix, where $x = I, U$ and $y = L, O, G, I, C$, is chosen at random from $[\frac{1}{l(I_{\alpha\beta}^{*t},x,y)}, l(I_{\alpha\beta}^{*t},x,y)]$ where $l(I_{\alpha\beta}^{*t},x,y)$ is the bound, and $I_{\alpha\beta}^{*t}$ is the intimacy.

The negotiation *strategy* is concerned with maintaining a working set of Options. If the set of options is empty then α will quit the negotiation. α perturbs the acceptance machinery (see Section 3) by deriving s from the $S_{\alpha\beta}^t$ matrix such as the value at the (I, O) position. In line with the comment in Footnote 4, in the early stages of the negotiation α may decide to inflate her opening Options. This is achieved by increasing the value of s in Equation 1. The following strategy uses the machinery described in Section 3. Fix h, g, s and c, set the Options to the empty set, let $D_s^t = \{\delta \mid \mathbb{P}^t(\mathrm{acc}(\alpha,\beta,\chi,\delta) > c\}$, then:

- repeat the following as many times as desired: add $\delta = \arg\max_x\{\mathbb{P}^t(\mathrm{acc}(\beta,\alpha,x)) \mid x \in D_s^t\}$ to Options, remove $\{y \in D_s^t \mid \mathrm{Sim}(y,\delta) < k\}$ for some k from D_s^t

By using $\mathbb{P}^t(\mathrm{acc}(\beta,\alpha,\delta))$ this strategy reacts to β's history of Propose and Reject utterances.

Negotiation *tactics* are concerned with selecting some Options and wrapping them in argumentation. Prior interactions with agent β will have produced an intimacy pattern expressed in the form of $\left(V_\alpha^{*t}, V_\beta^{*t}\right)$. Suppose that the relationship target is $(T_\alpha^{*t}, T_\beta^{*t})$. Following from Equation 3, α will want to achieve a *negotiation target*, $N_\beta(\Psi^t)$ such that: $v \cdot N_\beta(\Psi^t) + (1-v) \cdot V_\beta^{*t}$ is "a bit on the T_β^{*t} side of" V_β^{*t}:

$$N_\beta(\Psi^t) = \frac{v - \kappa}{v}V_\beta^{*t} \oplus \frac{\kappa}{v}T_\beta^{*t} \tag{4}$$

for small $\kappa \in [0, v]$ that represents α's desired *rate of development* for her relationship with β. $N_\beta(\Psi^t)$ is a 2×5 matrix containing variations in the LOGIC dimensions that α would like to reveal to β during Ψ^t (e.g. I'll pass a bit more information on options than usual, I'll be stronger in concessions on options, etc.). It is reasonable to expect β to progress towards her target at the same rate and $N_\alpha(\Psi^t)$ is calculated by replacing β by α in Equation 4. $N_\alpha(\Psi^t)$ is what α hopes to receive from β during Ψ^t. This gives a *negotiation balance target* of: $N_\alpha(\Psi^t) \ominus N_\beta(\Psi^t)$ that can be used as the foundation for reactive tactics by striving to maintain this balance across the LOGIC dimensions. A cautious tactic could use the balance to bound the response μ to each utterance μ' from β by the constraint: $V_\alpha(\mu') \ominus V_\beta(\mu) \approx S_{\alpha\beta}^t \otimes (N_\alpha(\Psi^t) \ominus N_\beta(\Psi^t))$, where $\otimes$ is element-by-element matrix multiplication, and $S_{\alpha\beta}^t$ is the stance. A less neurotic tactic could attempt to achieve the target negotiation balance over the anticipated complete dialogue. If a balance bound requires negative information revelation in one LOGIC category then α

will contribute nothing to it, and will leave this to the natural decay to the reputation D as described above.

6 Discussion

In this paper we have introduced a novel approach to negotiation that uses information and game-theoretical measures. It is grounded on business and psychological studies and introduces the concepts of *intimacy* and *balance* as key elements in understanding what is a negotiation strategy and tactic. Negotiation is understood as a dialogue that affect five basic dimensions: Legitimacy, Options, Goals, Independence, and Commitment. Each dialogical move produces a change in a 2×5 matrix that evaluates the dialogue along five information-based measures and five utility-based measures. The current Balance and intimacy levels and the desired, or target, levels are then used by the tactics to determine what to say next. We are currently exploring the use of this model as an extension of a currently widespread eProcurement software commercialised by a spin-off company of the laboratory of one of the authors.

References

1. Adams, J.S.: Inequity in social exchange. In: Berkowitz, L. (ed.) Advances in experimental social psychology, vol. 2, Academic Press, New York (1965)
2. Sondak, H., Neale, M.A., Pinkley, R.: The negotiated allocations of benefits and burdens: The impact of outcome valence, contribution, and relationship. Organizational Behaviour and Human Decision Processes, 249–260 (1995)
3. Lewicki, R.J., Saunders, D.M., Minton, J.W.: Essentials of Negotiation. McGraw Hill, New York (2001)
4. Sierra, C., Debenham, J.: Information-based agency. In: Proceedings of Twentieth International Joint Conference on Artificial Intelligence IJCAI-07, Hyderabad, India, pp. 1513–1518 (2007)
5. Sierra, C., Debenham, J.: Trust and honour in information-based agency. In: Stone, P., Weiss, G. (eds.) Proceedings Fifth International Conference on Autonomous Agents and Multi Agent Systems AAMAS-2006, Hakodate, Japan, pp. 1225–1232. ACM Press, New York (2006)
6. Fischer, R., Ury, W., Patton, B.: Getting to Yes: Negotiating agreements without giving in. Penguin Books (1995)
7. Valley, K.L., Neale, M.A., Mannix, E.A.: Friends, lovers, colleagues, strangers: The effects of relationships on the process and outcome of negotiations. In: Bies, R., Lewicki, R., Sheppard, B. (eds.) Research in Negotiation in Organizations, vol. 5, pp. 65–94. JAI Press, Greenwich (1995)
8. Arcos, J.L., Esteva, M., Noriega, P., Rodríguez, J.A., Sierra, C.: Environment engineering for multiagent systems. Journal on Engineering Applications of Artificial Intelligence 18 (2005)
9. Debenham, J.: Bargaining with information. In: Jennings, N., Sierra, C., Sonenberg, L., Tambe, M. (eds.) Proceedings Third International Conference on Autonomous Agents and Multi Agent Systems AAMAS-2004, pp. 664–671. ACM Press, New York (2004)
10. Sierra, C., Debenham, J.: An information-based model for trust. In: Dignum, F., Dignum, V., Koenig, S., Kraus, S., Singh, M., Wooldridge, M. (eds.) Proceedings Fourth International Conference on Autonomous Agents and Multi Agent Systems AAMAS-2005, Utrecht, The Netherlands, pp. 497–504. ACM Press, New York (2005)

Structural and Semantic Similarity Metrics for Web Service Matchmaking*

Akın Günay and Pınar Yolum

Department of Computer Engineering, Boğaziçi University,
Bebek 34342, Istanbul, Turkey
{akin.gunay,pinar.yolum}@boun.edu.tr

Abstract. Service matchmaking is the process of finding appropriate services for a given set of requirements. We present a novel service matchmaking approach based on the internal process of services. We model service internal processes using finite state machines and use various heuristics to find structural similarities between services. Further, we use a process ontology that captures the semantic relations between processes. This semantic information is then used to determine semantic similarities between processes and to compute match rates of services. We develop a case study to illustrate the benefits of using process-based matchmaking of services and to evaluate strengths of the different heuristics we propose.

1   Introduction

Web services are pieces of software that provide a functionality and can be invoked over the Web in a machine independent manner [1]. An important challenge in the usage of Web services is finding appropriate Web services for different service needs. Current standards, such as UDDI, only provide limited keyword search capabilities, which are insufficient to handle the requirements of the current users. With such protocols, users try to guess keywords that are relevant to their requests and Web services that advertise themselves with the same keywords are assumed to be good matches for each other.

Another influential trend in Web service matchmaking is that of input-output matching, where a service request is considered to match a Web service if the inputs and outputs are identical [2]. An enhancement to this approach is the addition of semantic information, where instead of identical matching partial matches are computed using the underlying semantic knowledge in the service descriptions [3][4]. In these semantic approaches, input-output fields are associated with semantic concepts represented in ontologies. The result of the matchmaking operation is a degree of semantic similarity, such as exact, plug-in and

---

* This research has been partially supported by Boğaziçi University Research Fund under grant BAP07A102 and the Scientific and Technological Research Council of Turkey by a CAREER Award under grant 105E073. The first author is supported by a Graduate Scholarship Program from the Scientific and Technological Research Council of Turkey.

G. Psaila and R. Wagner (Eds.): EC-Web 2007, LNCS 4655, pp. 129–138, 2007.

subsumes match. Exact match shows that the request and service interfaces match exactly to each other. Plug-in match shows that the service returns a more general output concept than the requested. Subsumes match shows that the service returns a more specific output concept than the requested.

Although input-output matching is easy to implement, it has two important drawbacks.

- *Granularity:* The results of the matchmaking is coarse-grained. That is, the matching services are associated only with some general similarity degrees (exact, plug-in, and so on) and we cannot further discriminate between services that have the same similarity degree. This level of granularity is unacceptable, especially when the number of matching services is large. A better matching approach should provide more precise matching information and should be able to rank the services based on this rating.
- *Precision:* The matchmaking algorithm should provide high precision; meaning that those Web services that are actually labeled as matching should be compatible with the requests. Most input-output matching techniques suffer from low precision, since they do not consider the internal processes of services while performing the matchmaking operation [5]. As a result, different services with identical interfaces are counted as good matches although they perform completely different tasks.

Accordingly, in this paper we propose a novel service matchmaking approach that uses internal process models of the services to achieve good precision and recall performances, while providing fine-grained similarity degrees. To achieve this, we propose to represent the underlying processes of Web services and the requests for Web services as finite state machines (FSMs). We perform matchmaking of requests to Web services by comparing the two FSMs and measuring their similarity using different metrics. Further, we associate each atomic process with a semantic concept represented in a process ontology and compare the semantic similarity of atomic processes during matchmaking. Our matching results combine the structural similarity with semantic similarity and provide fine-grained scores.

The rest of this paper is organized as follows. Section 2 shows our modeling of services as FSMs. Section 3 explains our matching approach in detail. Section 4 explains our case study and elaborates on our evaluations. Finally, Section 5 reviews some relevant literature and provides directions for future work.

2   FSM for Service Modeling

An FSM [6] is a formalism to capture the flow of processes. Using FSMs we can model the fundamental control structures (sequences, choices and loops) of process flows. A finite state machine is a 5-tuple $(Q, \Sigma, \delta, q_0, F)$, where Q is a finite set called *states*, Σ is a finite set called *alphabet*, $\delta : Q \times \Sigma \rightarrow Q$ is the *transition function*, $q_0 \in Q$ is the start state, and $F \subseteq Q$ is the set of final states. A is the language (the set of all strings) that machine M recognizes and we show this by A $L(M)$.

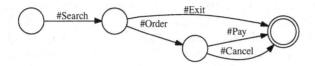

Fig. 1. A sample e-commerce service flow

In our approach, each element of the alphabet Σ represents an atomic process (or simply a process) of the modeled service. We use the states in Q to investigate the order of flow. Each transition function captures a process can be performed in the internal flow of the service. The start state is the entry point of the service and final states are the termination points. Each string in the language A is a sequence of consecutive processes that corresponds to a service flow.

Figure 1 shows an example of an e-commerce service's internal process flow modeled as an FSM. From the entry point (the start state) of the e-commerce service, the first process that we can perform is the #Search process, which returns some item(s) according to the requesters search criteria. According to the result of the #Search process, the requester may choose to buy the resulting item by performing #Order process or may end the interaction with the service by #Exit process. If the requester selects #Order process in the previous step, she may continue with the payment by performing the #Pay process or quit the service by canceling the order via #Cancel process. Figure 2 shows all possible flow sequences for the FSM presented in Figure 1.

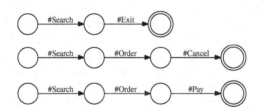

Fig. 2. All process flow sequences (language) generated by the FSM in Figure 1

3   Matchmaking Approach

Given a service request, it is necessary to match it to a set of services from a pool of available services. To determine the similarity of a request to a particular Web service, we compute its *structural* and *semantic* similarity.

3.1   Structural Similarity

Since we use FSMs to model services, our first option to perform matchmaking is the use of formal definition of FSM equality. According to this definition, two

FSMs are equivalent to each other if they recognize the same language. In our context this definition is too restrictive because the equality definition supports only services that exactly match each other. However, we are also interested in finding partially matching services as well as determining their match rates.

Because of these reasons we need a more flexible mechanism that can find partially matching services in addition to exact matches. Further, the matchmaking mechanism should assign a numeric similarity value to each service in order to differentiate between services. To achieve these goals, we use the following approach instead of the equality definition of the FSM.

Our approach is made up of the following steps:

1. Generate all possible flow sequences (all strings in the language) for the service and request using the associated FSMs.
2. Compare each sequence of the request against all the sequences of the service.
3. For each sequence comparison, compute a similarity value between the compared request and service sequence using a heuristic function (explained below).
4. Select the sequence with highest similarity value from service sequences.

Optionally, we can compute the scores using different heuristics and combine the results as the normalized sum of the selected pairs. We perform the above procedure for all the available services and sort the services in decreasing order according to their overall similarity values and return the top n percent as result.

To generate all flow sequences of the FSM, we expand the FSM from the start state up to the final states. This procedure is simple if the FSM structure is an acyclic directed graph. But if we introduce loops to the FSM, the graph turns to a cyclic directed graph where the number of possible sequences is infinite. To handle this case we modify our expansion algorithm so that it can detect loops and stop expanding a sequence when the same loop occurs more than once.

To compute the structural similarities we use four heuristics. These heuristics are *common process count* (CPC), *longest common substring* (LCStr), *longest common subsequence* (LCSeq) and *edit distance* (ED). Below, s_1 is a service request and s_2 is a Web service.

Common Process Count Heuristic. The common process count (CPC) heuristic calculates the number of processes that appear in the service request and in given Web service sequences, without regard to the order of processes and normalizes the count with the total number of processes in the sequences. The underlying intuition is that when the number of common processes for the services increase, the two sequences are more similar to each other. The similarity between s_1 and s_2 is computed as follows where N_p is the number of common processes and N_{s_i} is the number of processes in s_i.

$$sim(s1, s2, CPC) = \frac{2N_p}{N_{s_1} + N_{s_2}} \tag{1}$$

Longest Common Substring Heuristic. The LCStr heuristic [7] finds the longest common contiguous substring of two strings. In our context, this corresponds to the number of contiguous processes between the request and a Web service. Formally, using LCStr heuristic, similarity between s_1 and s_2 is computed and normalized as follows:

$$sim(s1, s2, LCStr) = \frac{length(LCStr(s_1, s_2))}{length(s_1)} \tag{2}$$

where $length(s)$ is a function that returns the length of the string s. $LCStr(s_i, s_j)$ is a function that returns the common longest substring of s_i and s_j.

Longest Common Subsequence Heuristic. The LCSeq heuristic [7] finds the longest common subsequence (may not be contiguous) of two strings. In our context, this corresponds to finding the number of common processes between a request and a Web service without considering the contigiousness. Formally, using LCSeq heuristic, similarity of s_1 and s_2 is computed and normalized as follows:

$$sim(s1, s2, LCSeq) = \frac{length(LCSeq(s_1, s_2))}{length(s_1)} \tag{3}$$

where $length(s)$ is a function that returns the length of the string s. $LCSeq(s_i, s_j)$ is a function that returns the common longest subsequence of s_i and s_j.

Edit Distance (Levenshtein) Heuristic. Edit distance [7] is the minimum number of operations needed to transform one string into another, where an operation is an insertion, deletion, or substitution of a single character. In our context the strings are the flow sequences and the characters are the processes. We formally calculate the similarity of s_1 and s_2 as follows:

$$sim(s1, s2, ED) = 1 - \frac{ED(s_1, s_2)}{Max(length(s1), length(s2))} \tag{4}$$

where $ED(s_1, s_2)$ is the function that computes the edit distance of the sequences and $Max(l_1, l_2)$ is the function that returns the maximum of two integers.

3.2 Semantic Similarity

Our structural similarity method works only on the syntactic level and do not take the underlying semantics into account. That is, if two Web services have entirely different atomic processes, the structural similarity of these services will be zero. However, if the atomic processes are related to each other, then the two Web services can still be considered similar. For example, one Web service may use allow *payByCreditCard* process whereas a second Web service may use *payByCash* process to handle payment. For a structural similarity metric, these are two totally different processes. However, one can easily see that they are two variations of the same process and hence may substitute each other in many settings. To consider such subtleties, the relations between the processes should

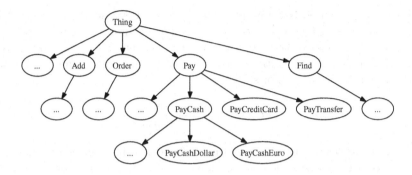

Fig. 3. A part of the process ontology for e-commerce domain

be captured. We do this using an ontology of processes that represents the meanings and relations of the processes. Each concept in this ontology corresponds to an atomic process. The children of a concept are the specializations of the process. In this ontology we assume that a more general process can perform all the tasks performed by its more specialized processes (sub-concepts). Figure 3 shows a part of our process ontology for the e-commerce domain.

To compute the semantic similarity of two concepts we develop a new semantic similarity metric called *semantic cover rate* (SCR). Using the SCR semantic similarity of two concept c_1 and c_2 is calculated as follows:

$$SCR(c_1, c_2) = \begin{cases} 1, & \text{if } c_1 \supseteq c_2 \\ \theta^{\|c_1, c_2\|}, & \text{if } c_1 \subset c_2 \\ \gamma^{\|R, c_1\|}, & \text{if } c_1 \not\supseteq c_2 \text{ and } c_1 \not\subseteq c_2 \text{ and } R \supset c_1 \text{ and } R \supset c_2 \\ 0, & \text{otherwise} \end{cases} \quad (5)$$

where θ and γ are two control parameters in the range $[0,1]$ and $\|c_i, c_j\|$ is the arc distance in the ontology between the concepts c_i and c_j. It is important to note that $SCR(c_i, c_j) \neq SCR(c_j, c_i)$.

In the first case, the request is a subset of a given Web service. Since we assume that a general process can perform all the tasks performed by its more specialized processes, we assign a SCR value of 1 for this case. However, if the service provides a more specific process than the requested process, the service process can handle only some of the needs of the requested process and therefore we assign a SCR value smaller than 1 by using θ parameter. In the third case, the considered concepts are siblings. Although in this case there is no super or sub-concept relation between the two concepts, since these concepts have a common root they are still related to each other. However, in this case the resulting SCR value must be smaller than the previous case, since the relation between the concepts are weaker. We achieve this effect by assigning γ a smaller value than θ. In the last case, since there is no relation between the concepts, we assign zero as SCR value.

4   Evaluation

The following case study presents a service request and six services, which match to the request by different degrees.

4.1   Case Study Setup

Both the request and the services consist of only one flow sequence. We ignore the case of multiple choices and loops in our case study, since these control structures are converted into single sequences during the matchmaking process and do not have any effect on the computation of the similarity values.

Request: #ConnectNonSecure → #SearchBook → #AddCartBook →
#OrderCartBook → #Authenticate → #PayCreditCard. This request is for a book shopping service. It requires a nonsecure connection and basic shopping functionalities like search and order. For payment it requires a credit card payment capability using any kind of authentication mechanism.

Service-1: #ConnectNonSecure → #SearchBook → #Authenticate →
#AddCartBook. This service provides search and cart functionalities but it is not possible to order the card or make any payments. This service is a bad match for the request, since it does not provide most of the fundamental functionalities of the request like ordering the cart and payment.

Service-2: #ConnectSecure → #SearchBook → #AddCartBook →
#OrderCartBook → #AuthenticateHTTPS → #PayMoneyTransfer. This service provides all the fundamental functionalities required by the request. However some processes that provide these functionalities are different than the requested service. For example the service provides the payment functionality by money transfer, which is different than the requested credit card payment. This service is an average match for the request, since it provides all the functionality to buy a book, but with some different processes.

Service-3: #ConnectNonSecure → #SearchDVDBasicTitle → #AddCartDVD → #OrderCartDVD → #Authenticate → #PayCreditCard. This service is to buy a DVD but not a book. Therefore this is not a good match for the request.

Service-4: #Connect → #Search → #AddCart → #OrderCart → #CancelOrder → #Authenticate → #PayCreditCard. This service is a general service where it is possible to search, order and pay for any consumable item. It is also possible to cancel a given order, which is not required by the request. Since this service covers also purchase of a book, it is a good match for the request.

Service-5: #ConnectNonSecure → #SearchBookBasicTitle → #AddCartBook → #OrderCartBook → #AuthenticateSSH → #PayCreditCardVisa. This service provides more specific functionalities than the requested service. For example it accepts search only by book title, where the request looks for a service that can provide any type of search capabilities. This service can be accepted as an average match, since it mostly provides the requested functionalities.

Service-6:   #ConnectNonSecure → #Authenticate → #SearchBook → #AddCartBook → #OrderCartBook → #PayCreditCard. This service provides exactly the requested functionality but in a different order. The authentication is performed after the connection instead of before payment, which causes shift of all processes in the flow order.

Overall, one would expect services 4 and 6 to be the better matches for the request. Next, we study the performance of our approach.

4.2   Results

To measure the individual performance of each structural heuristic first we compute the similarity between the request and all services for each heuristic separately. Than we integrate the SCR heuristic to each structural heuristic and perform the same procedure to observe the effect of the semantic knowledge. In addition to the individual heuristics, to observe the effect of the combination of different heuristics, we also compute the weighted linear combination of all the heuristics.

Table 1 presents our results. Each column shows the computed similarities between the request and each service using the associated heuristic. We add the letter S to the beginning of the heuristic names to indicate that they use SCR metric in addition to the structural similarity. The column named as Comb shows the results that we obtain by the linear combination of all heuristics (with equal weights). In the SCR computations we take θ as 0.75 and γ as 0.5. Considering

Table 1. Matchmaking results of each heuristic

	CPC	SCPC	LCStr	SLCStr	LCSeq	SLCSeq	ED	SED	Comb	SComb
Serv-1	0.60	0.94	0.50	0.33	0.75	0.50	0.33	0.46	0.55	0.56
Serv-2	0.50	0.79	0.50	0.79	0.50	0.79	0.50	0.79	0.50	0.79
Serv-3	0.50	0.69	0.33	0.69	0.50	0.69	0.50	0.69	0.46	0.69
Serv-4	0.15	0.86	0.14	0.67	0.14	1.00	0.14	0.86	0.15	0.85
Serv-5	0.50	0.84	0.33	0.84	0.50	0.84	0.50	0.84	0.46	0.84
Serv-6	1.00	1.00	0.50	0.50	0.83	0.83	0.67	0.67	0.75	0.75

these results, we observe the following:

- CPC is particularly useful when the request and the service have the same functionality but different process flows like in the case of Service-6. CPC can also successfully differentiate the unrelated services such as Service-3. However, it cannot detect that Service-1 is not a good match since CPC does not consider the difference between the number of processes in the request and the number of processes in the service.

- In general LCStr shows the worst performance compared to the other heuristics, since it is strictly dependent on the order of the process flow. It is only successful in the case of Service-1, where the service provides less functionality and therefore the longest common substring between the service and request is short.
- LCSeq is especially successful if the service covers the request and provides some additional functionality like in the case of Service-4. It can also successfully detect the two poor matches Service-1 and Service-3.
- ED can successfully differentiate between good and poor matches. The only exception occurs when flow orders are different like in the case of Service-6.
- Considering both the structural and semantic similarity gives more accurate results compared to the use of structural similarity alone.
- An intuitive approach to improve the quality of results is to use a combination of the heuristics instead of using them individually. The results obtained by the Comb prove this idea.

5 Discussion

In this study we propose a new semantic matchmaking approach for Web services that is based on the internal process flows of the services. Our approach combines structural similarity heuristics with a semantic similarity metric based on ontologies. To determine the similarity of two services, structural heuristics compare the individual atomic processes that are involved by the service flows. In this comparison if two atomic processes are identical, structural heuristics assign 1 as the similarity value and 0 otherwise. This approach restricts us to exactly matching processes. To be able to discover partially matching processes, we relax the similarity of non-identical processes into the range [0,1]. Our approach requires that both services and requests are modeled as FSMs by developers and users. Especially for the users developing a FSM of a service might be complex. However, in real life applications of this approach , we can count on the existence of support tools that can help users define their requests as FSMs.

Klusch et al. [5] extend the signature matching approach by using syntactic matching techniques from information retrieval on service descriptions in order to improve granularity and precision performance. Although this approach provides a mechanism to associate each service with a numeric similarity value, it still suffers from low precision. Additionally, Dong et al. [8] state that syntax based information retrieval techniques are not efficient for Web service matching, since in most of the real world situations service descriptions do not contain enough textual data that is required by syntactic methods to work properly.

There are other approaches for matchmaking that use process modeling techniques. Klein and Bernstein [9] propose an indexing mechanism to create a hierarchical ontology of process models and develop a query language to perform matching on the created ontology. Wombacher et al. [10] propose another approach which also use FSA to model Web services. Different than our approach they concentrate on the syntactic level matching of FSA. They do not consider any ontologies for semantics or do not work on a rank mechanism or partial matching. Addition to discovery processes modeling, FSA is also studied for other Web service issues like composition. Berardi *et al.* [11] propose a service composition approach that can work with time constraints.

In our future work, we plan to optimize SCR performance by deciding on the δ and γ values at run time. This will enable us to achieve more accurate matches. Another interesting direction is the investigation of fuzzy is-a relations to represent the domain. The weights again may be learned at runtime to improve personalized match performance.

References

1. Singh, M.P., Huhns, M.N.: Service-Oriented Computing: Semantics, Processes, Agents. John Wiley & Sons, Chichester (2005)
2. Zaremski, A.M., Wing, J.M.: Specification matching of software components. In: SIGSOFT '95. Proceedings of the 3rd ACM SIGSOFT Symposium on Foundations of Software Engineering, New York, NY, USA, pp. 6–17. ACM Press, New York, NY, USA (1995)
3. Paolucci, M., Kawamura, T., Payne, T.R., Sycara, K.P.: Semantic matching of Web services capabilities. In: Proceedings of the First International Semantic Web Conference, pp. 333–347. Springer, Heidelberg (2002)
4. Sycara, K., Widoff, S., Klusch, M., Lu, J.: Larks: Dynamic matchmaking among heterogeneous software agents in cyberspace. Autonomous Agents and Multi-Agent Systems 5(2), 173–203 (2002)
5. Klusch, M., Fries, B., Sycara, K.: Automated semantic web service discovery with owls-mx. In: AAMAS '06. Proceedings of the Fifth International Joint Conference on Autonomous Agents and Multiagent Systems, New York, NY, USA, pp. 915–922. ACM Press, New York (2006)
6. Sipser, M.: Introduction to the Theory of Computation. 2nd edn., Course Technology (2005)
7. Gusfield, D.: Algorithms on Strings, Trees, and Sequences: Computer Science and Computational Biology. Cambridge University Press, Cambridge (1997)
8. Dong, X., Halevy, A.Y., Madhavan, J., Nemes, E., Zhang, J.: Simlarity search for web services. In: VLDB, pp. 372–383 (2004)
9. Klein, M., Bernstein, A.: Toward high-precision service retrieval. IEEE Internet Computing 8(1), 30–36 (2004)
10. Wombacher, A., Fankhauser, P., Mahleko, B., Neuhold, E.: Matchmaking for business processes based on conjunctive finite state automata. International Journal of Business Process Integration and Management 1(1), 3–11 (2005)
11. Berardi, D., Giacomo, G.D., Lenzerini, M., Mecella, M., Calvanese, D.: Synthesis of underspecified composite e-services based on automated reasoning. In: Proceedings of the 2nd International Conference on Service Oriented Computing, New York, NY, USA, pp. 105–114. ACM Press, New York (2004)

Providing Methodological Support to Incorporate Presentation Properties in the Development of Web Services*

Marta Ruiz, Pedro Valderas, and Vicente Pelechano

Universidad Politécnica de Valencia
Camí de Vera s/n, Valencia-46022, España
{mruiz, pvalderas, pele}@dsic.upv.es

Abstract. Nowadays, Web services provide only raw data. However, Web service consumers need Web services with new features that allow them to obtain data with specific presentation properties. In this paper, we present an approach to develop Web services that incorporate presentation properties from a methodological point of view. To do this, we based on the models proposed by the model-driven development method OO-Method / OOWS. To document our approach, we apply our ideas to a real case study of a Web application to manage University Research Groups.

1 Introduction

A second generation of web-based services that emphasize online collaboration and sharing among users is currently rising. It is known as Web 2.0. Although the meaning of "Web 2.0" is still under debate, in the opening talk of the first Web 2.0 conference, Tim O'Reilly and John Battelle summarized the key principles of Web 2.0 applications. One of these principles was "innovation in assembly of systems and sites composed by pulling together features from distributed and independent developers."

This innovation in the assembly of distributed features can be appreciated in the arising of new technology concepts such as Mashups [5] or Portlets [1]. A Mashups is a website or application that combines content from more than one source into an integrated experience. Portlets are pluggable user interface components that are managed and displayed in a web portal. The user can personalize the content, appearance, and position of the Portlets. Related to the Portlets concept, there exists the Web Services for Remote Portlets (WSRP) protocol whose purpose is to provide a Web service standard that allows for the "plug-n-play" of remote running Portlets from disparate. For instance, Figure 1 shows an example of the same Portlet (www.accuweather.com) accessed from three different portals.

From a Service Oriented Architecture (SOA) perspective a Portlet that is remotely accessed can be considered to be a Web services that provide a fragment of markup

---

* This work has been developed with the support of MEC under the project DESTINO TIN2004-03534 and cofinanced by FEDER.

G. Psaila and R. Wagner (Eds.): EC-Web 2007, LNCS 4655, pp. 139–148, 2007.

code that shows information with a specific look and feel. This code is prepared to be directly interpreted by a Web browser in order to allow users interacting with it.

However, little methodological support is provided to develop this kind of services. From a Web Engineering perspective there are a lot of methods which provide support for developing Web applications from conceptual models. Some of these methods such as OOHDM [13] and UMLGuide [4] provide abstract mechanisms to include Web services calls in the definition of their conceptual models. However, no methodological support is provided to develop Web services from their conceptual models and then publish them. Other methods like WebML [2] provides more specific support to the development of Web services however it does not provide methodological support to define them. In a previous work [10], we introduce a methodological guide that allows us to automatically develop Web services from the OO-Method / OOWS [8] models. However, the generated Web services do not consider aspects related with the presentation of information.

Fig. 1. AccuWeather's Portlets

In this paper, we extend our previous work by introducing new guidelines that allow us to automatically generate "presentation Web services" from the OO-Method/ OOWS conceptual models. These services provide the same behaviour than a Portlet does. In the same way, these services provide the proper code in order to be used as part of a Mashup. The main advantage of use our Web services instead of other solutions are: (1) they are automatically generated following a proper methodological guide; (2) they can be easily managed and extended from the end-user point of view.

The rest of the paper is organized as follows: an overview of the OO-Method/OOWS development process is shown in Section 2. Section 3 describes the main extensions to enrich the Web service approach in order to include presentation properties. Finally, conclusions are presented in section 4. The Web application that is used to manage our research group (Http://oomethod.dsic.upv.es) has been taken as a case study to clearly map the new concepts and abstraction mechanisms and their implementation in a real web environment.

2   An Overview of the Development Process

In this section, we introduce a methodological process to obtain Web services from conceptual models. To present this process we use the notation and terms defined in the Software Process Engineering Metamodel (SPEM) proposed by the OMG [14]. SPEM is a meta-model for defining software engineering process models and their components. According to the SPEM, process can be defined from a set of activities. An *Activity* is a piece of work performed by one *ProcessRole* in order to obtain a *WorkProduct*. A *Discipline* partitions the Activities within a process according to a common "theme".

Figure 2 presents the Disciplines that define our process. There are four disciplines: (1) Requirements elicitation, (2) Conceptual Modelling, (3) Code Generation and (4) Web services. Next, we present the *Activities*, *ProcessRoles* and *WorkProducts* that are included in each discipline.

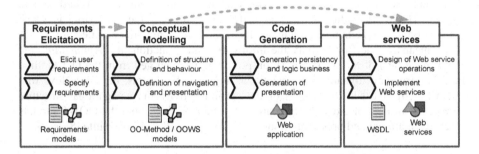

Fig. 2. Process followed to design and implement Web services

2.1 Requirements Elicitation

The Requirements Elicitation discipline includes those activities that are related to the elicitation of the user's requirements. These activities are two: (1) elicit the user's requirements and (2) specify those requirements. The ProcessRole which perform these activities is Analyst. The WorkProduct that is obtained is a requirements model. In order to obtain detailed information about the definition of this requirements model see [15].

2.2 Conceptual Modelling

This discipline includes those activities related to the modelling of a Web application. We distinguish two main activities: (1) definition of structure and behaviour and (2) definition of navigation and presentation. These activities are performed by Modeller. The WorkProduct that is obtained is a set of models which captures the structure, behaviour, navigation and presentation of a Web application at conceptual level.

These activities are performed by following the development process of the OO-Method / OOWS method. OO-Method [7] is an engineering method that provides methodological support for development of traditional software (no-Web). OOWS [8] is its extension for the development of Web applications.

In order to develop a software system OO-Method proposes the construction of three models: (1) the *structural model* which describes the static structure; (2) the *dynamic* and (3) *functional* models which are both used to describe the system behaviour.

OOWS introduces three new models in order to capture new aspects encouraged in the development of Web applications: (4) the *user model* which describe the different types of users that can connect the Web application; (5) the *navigational model* which describes the navigational structure; and (6) the *presentation model* to specify presentation requirements of a Web application.

2.3 Code Generation

This discipline includes those activities that are related to the generation of code from conceptual models. We distinguish two main activities: (1) generation *persistency* and *logic business* and (2) generation of *presentation*. The ProcessRoles which perform these activities are the case tools which support both methods OO-Method and OOWS. The WorkProduct that is obtained after performing these activities is a web application prototype that is implemented by following a three-tier architecture.

By means of the first activity the information and functionality (*Application* and *Persistence* tier) of the web application prototype is generated from the OO-Method models (structural, dynamic and functional models). To do this, the OlivaNova tool (ONME) [6] is used. By means of the second activity the navigational structure (*Presentation* tier) of the web application is generated from both the OOWS models (user, navigational and presentation models). To do this, the OOWS Suite [8] is used.

2.4 Publication of Functionality Throughout Web Services

This discipline includes those activities that allow us to publish the functionality defined in the OO-Method / OOWS models throughout Web services. To do this, a methodological guide has been defined in order to allow us to automatically define the Web services and then implement them (see [10] and [11]). Thus, we distinguish two main activities: (1) Design of Web service operations and (2) implement a fully operative Web service. The ProcessRole of these activities is a tool which supports the Web service methodology. The WorkProduct that is obtained after performing these activities are a WSDL specification and a Web service implementation.

The proposed methodological guide allows us to automatically obtain Web services operations from OO-Method / OOWS models. These Web services provide operations that implement the requirements of a Web application in a SOA. The models that are taken as source in order to obtain the Web services are: the *Requirement Models* defined in the Requirements Elicitation discipline and the *Structural Model*, *User Model* and the *Navigational Model* defined in the conceptual modelling discipline.

The operations that are automatically generated from these models use the functionality that is generated by ONME. These operations allow us to publish the following functionality (see Figure 3): (a) *Information Management*: Operations which allow the management of the population of objects which define the information system. These operations depend on the web application domain. For instance, in the running example, operation such as `newPublication`, `addMemberToPublication` or `newMember`, are defined. (b) *Information retrieval*: Operations which allow the retrieval of information. They depend on the web application domain. For instance, in the running example, operations such as `retrieveMembers` or `searchMember` are defined. (c) *User identification and management*. Operations which allow the management of users. These operations do not depend on the Web application domain. They are the same for every Web application. Some of these operations are: `loginUser`, `logoutUser`, `changeRol`, `remindPassword`, `newUser` or `modifyUser`.

Figure 3 shows an example of the operations that are published for the running example. Figure 3A shows the WSDL of the Web service generated for the running example and the Figure 3B shows the detail of the `retrieveMembers` operation.

Fig. 3. Some operations of the Web service generated to the OO-Method Web site

These operations just return a structured data in a XML document without presentation properties. In the next section, we present an extension of our methodology in order to incorporate presentation characteristics to the returned data. This allows us to provide a functionality equivalent to the Portlets one throughout Web services. We also improve the Web service in order to facilitate the mashing up of its operations to provide fully operative Web pages.

3   An Extension of the Methodological Guide to Provide Presentation

In this section, an extension of our methodological guide is introduced. This extension allows us to automatically generate Web services operations which retrieve data with specific presentation properties. These presentation properties are extracted from the OOWS presentation model (see section 2.2). Thus, we present first the presentation model in detail. Next, we introduce the proposed extensions.

3.1   The OOWS Presentation Model

Modellers must specify presentational requirements of web applications by using the presentation model once the navigational model is built. The presentation model is strongly based on the navigational model because it uses the conceptual primitives of the navigational model in order to define the presentation properties.

The navigational model is defined as direct graph whose nodes represent navigational contexts and whose arcs represent navigational links. A navigational context defines a view over the class diagram which allows us to indicate the information that must be provided to the user as well as the operations that the user can access. Figure 4A shows the *Member* navigational context. This context provides detailed information about members and the research group to which they belong.

In order to define the presentation model we must take as source the classes and relationships defined in each navigational context and then attach them presentation patterns. Figure 4B shows the presentation patterns associated to the context in figure 4A. The basic presentation patterns that can be defined are: (a) *Information Paging*. This pattern allows defining information "scrolling". All the instances are "broken" into "logical blocks", so that only one block is visible at a time. Mechanisms to move forward or backward are provided. The required information is: (1) cardinality represents the number of instances that make a block; and (2) access mode (sequential or random). (b) *Ordering Criteria*. This pattern defines a class population ordering (ASCendant or DESCendant). (c) *Information Layout*. OOWS provides 4 basic layout patterns: register, tabular, master-detail and tree.

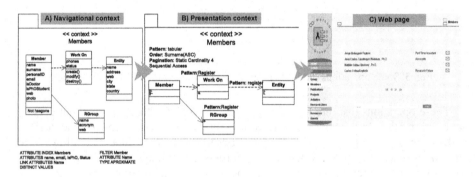

Fig. 4. The OOWS Discipline, where the Presentation Model is highlighted

According to the requirements specified in the figure 4B, the list of members provided by the context must be showed in a tabular format (such a list of elements). This list is grouped in blocks of four members and the surname of each member is used to order them in an increase way. Figure 4C show a Web page which provides the information defined in the navigational context in figure 4A with the presentation properties defined in the presentation model in Figure 4B.

3.2   Extending Web Services

In this section, we introduce the new operations that are incorporated to the Web Service in order to support the functionality that provide Portlets or Mashups.

3.2.1   An Extension to Retrieve Data with Presentation Properties

The first extension that we propose is the definition of a new operation that allows us to incorporate presentation properties to the data that is obtained by the operations presented in Section 2.4.

The signature of this operation is: `presentationInfo` `(operation,` `operationParameters[,paging,order,layout])`. This operation calls to the operation indicated by `operation` with the parameters indicated by `operationParameters`. The optional attributes `paging`, `order` and `layout` indicates the presentation patterns in which the information must be shown.

If the user does not give any value of these presentation parameters, the result is returned with the presentation patterns defined by the modellers in the presentation model. To better understand this aspect we must consider that operations commented in Section 2.4 are defined from the views over the class diagram that define each navigational context [8]. For instance, the operation `retrieveMembers` retrieve the information defined in the context Members (see Figure 4A). If we directly call to this operation we obtain a XML code that is shown in a Web browser such as we can see in Figure 5A. However, if we call to this operation throughout the operation `presentationInfo` we obtain a code that is shown in a Web browser such as we can see in Figure 5B. This code is the same as Figure 5A however the presentation properties defined in the presentation model have been associated.

Fig. 5. A) XML document returned by `searchMember`; B) the result of `presentation-Info (searchMember, parameters)`

In order to define and associate presentation properties to the XML code, a strategy based on CSS [3] templates is used. If we call an operation throughout `presentationInfo`, the properties defined in the presentation model for this operation are accessed and then defined in a CSS template. Next, `presentationInfo` call to the operation and modify the obtained XML code in order to associate it to the CSS template. Finally, `presentationInfo` sends to the Web browser the modified XML code and the CSS template.

As we have commented above, it is possible that consumers of the Web service operation `presentationInfo` indicate the presentation patterns to be applied by means of the parameters paging, order and layout. If these parameters are indicated, then the properties of the presentation model are not taking into account. We have provided this possibility in order to facilitate Web service consumers to adapt the presentation properties to their preferences.

3.2.2 An Extension to Obtain a Web page

The second extension of the Web service methodology defines an operation for each context which fully implements the equivalent Web page. For instance, we automatically generate an operation `presentationMembers` which create the Web page equivalent to the context Members in figure 4A.

In order to generate this type of operations, a Web page is considered to be an aggregation of a set of logical content areas (see [12] for more detailed information). Some representative examples of these areas are:

- *Information area* (see Figure 6A, zone number as 1): presents the data and the functionality that users can access. This area can be implemented using those operations that give support to retrieve the information.
- *Navigation area* (see zone 2): provides links to the web pages that are available to users. This area can be implemented using those operations that give support to the navigational links.
- *Access-Structure area* (see zone 3): provides users with mechanisms such as search engines or information indexes, which facilitates the access to the information. This area can be implemented using those operations that implements search or index operations.

Then, previously to generate operations such as `presentationMembers` we generate operations that implement the commented areas. Each of this operation is implemented by calling the operations introduced in Section 2.4. For instance, we generate an operation `InformationAreaMembers` which call to the operation `retrieveMembers`. In order to incorporate presentation properties it is called throughout the operation `presentationInfo`.

Figure 6 shows the Web page created as a result of the `presentation-Members` operation. The information retrieved from the Members view is showed in register, in groups of four and ordered by surname increasing. In addition, it shows links to those pages reachable since this context.

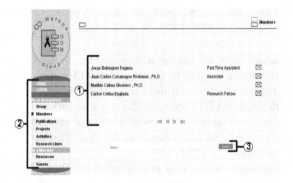

Fig. 6. The result of `presentationMembers` operation

Thus, the operation `presentationMembers` is implemented by calling to the different operations that implement its areas. `PresentationMembers` is implemented by calling to the operations `InformationAreaMembers`, `NavigationAreaMembers` and `AccessStructureAreaMembers`. In the same way, these operations are implemented by calling to the operations presented in Section 2.5. As commented above, these calls are performed throughout the operation presentationInfo.

The operations that implement areas can be reused. For instance, it is possible that users want to create a Web page that provides information about members and also about publications. Then users can create this pages by calling to the operations InformationAreaMembers and the operation InformationAreaPublications (obtained form the contexts Publications that is omitted due to space constraints). Thus, a Web page can be built by aggregating operations published in the Web service with minimal development effort (such as Mashups do) as commented in section 3.2. Figure 7 shows an example of this. Figure 7A shows the code which implement a Web page from callings to the area based operations. As we can see there are a call to the operation InformationAreaMembers and other call to the operation InformationAreaPublications. Figure 7B shows the resulting implementation.

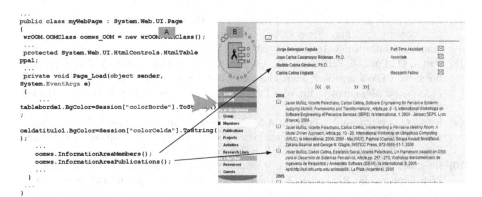

Fig. 7. The implementation of the InformationArea operations

4 Conclusions and Future Work

This article has presented an approach to introduce presentation properties to the Web services development. This approach takes as source to extract the presentation properties the OO-Method / OOWS presentation model.

Three Web sites have been successfully developed using the methodology presented in this paper: the OO-Method Group Web Site (http://oomethod.dsic.upv.es), the intranet management system for the General Library of the Technical University of Valencia and the DSIC Department Web Site (http://www.dsic.upv.es).

Our work opens up interesting venues for further research. The extension proposed in this article has as another objective to generate reusable Web functionality that provides relevant information (content and presentation) to potential consumer [9]. We are also working on integrating this approach on the OlivaNova Model Execution tool [6].

References

[1] BEA Web Logic Portal. Portlet Developer's Guide http://edocs.beasys.com/wlac/portals/docs/portlet.html

[2] Brambilla, M., Ceri, S., Comai, S., Fraternali, P., CASE,: A CASE tool for modelling and automatically generating web service-enabled applications. Int. J. Web Engineering and Technology 2(4) (2006)

[3] Cascading Style Sheets. W3C (2005), http://www.w3.org/Style/CSS/

[4] Dolog, P.: Model-driven navigation design for semantic web applications with the UML-guide. In: Matera, M., Comai, S., (eds.) Engineering Advanced Web Applications (2004)

[5] Jhingran, A.: Enterprise information mashups: integrating information, simply. In: Dayal, U., Whang, K., Lomet, D., Alonso, G., Lohman, G., Kersten, M., Cha, S.K., Kim, Y. (eds.) Proceedings of the 32nd international Conference on Very Large Data Bases, Very Large Data Bases. VLDB Endowment, Seoul, Korea, September 12 - 15, 2006, vol. 32, pp. 3–4 (2006)

[6] OlivaNova Model Execution System. CARE Technologies. Retrieve (January 2006), from: http://www.care-t.com

[7] Pastor, O., Gomez, J., Insfran, E., Pelechano, V.: The OO-Method approach for information systems modelling: from Object-Oriented conceptual modeling to automated programming. Information Systems 26 (2001)

[8] Pastor, O., Fons, J., Pelechano, V., Abrahão, S.: Conceptual modelling of Web applications: the OOWS approach, book chapter in Web Engineering - Theory and Practice of Metrics and Measurement for Web Development. In: Mendes, E. (ed.)Springer, Heidelberg (2005)

[9] Rojas, G., Pelechano, V., Fons, J.: A Model-Driven Approach to include Adaptive Navigational Techniques in Web Applications. In: International Workshop on Web Oriented Software Technologies - IWWOST, Porto, Portugal (2005)

[10] Ruiz, M., Valderas, P., Pelechano, V.: Applying a Web Engineering Method to Design Web Services. In: Benatallah, B., Casati, F., Traverso, P. (eds.) ICSOC 2005. LNCS, vol. 3826, Springer, Heidelberg (2005)

[11] Ruiz, M., Pelechano, V., Pastor, O.: Designing Web Services for Supporting User Tasks: A Model Driven Approach. In: Roddick, J.F., Benjamins, V.R., Si-Saïd Cherfi, S., Chiang, R., Claramunt, C., Elmasri, R., Grandi, F., Han, H., Hepp, M., Lytras, M., Mišić, V.B., Poels, G., Song, I.-Y., Trujillo, J., Vangenot, C. (eds.) Advances in Conceptual Modeling - Theory and Practice. LNCS, vol. 4231, Springer, Heidelberg (2006)

[12] Valderas, P., Pelechado, V., Pastor, O.: Introducing Graphic Designers in a Web Development Process. In: 19th International Conference on Advanced Information Systems Engineering (CAiSE'07) - Trondheim, Norway (2007)

[13] Schwabe, D., Rossi, G., Barbosa, D.J.: Systematic hypermedia application design with OOHDM. In: Proc. ACM Conference on Hypertext, pp.166 (1996)

[14] Software Process Engineering Metamodel, version 1.1. Object Management Group. http://www.omg.org/technology/documents/formal/spem.htm

[15] Valderas, P., Fons, J., Pelechano, V.: Developing E-Commerce Application from Task-Based Descriptions. In: Bauknecht, K., Pröll, B., Werthner, H. (eds.) EC-Web 2005. LNCS, vol. 3590, Springer, Heidelberg (2005)

A Model of IT Evaluation Management: Organizational Characteristics, IT Evaluation Methodologies, and B2BEC Benefits

Chad Lin[1] and Yu-An Huang[2]

[1] CBRCC, Curtin University of Technology, Australia
elin123au@yahoo.com.au
[2] Department of International Business Studies, National Chi Nan University, Taiwan
yahuang788@gmail.com

Abstract. Large organizations have invested substantial financial resources in information technology (IT) over the last few decades. However, many organizations have discovered that they have not yet fully reaped the B2BEC benefits from their IT investments. A model of IT evaluation management is proposed in this paper to examine: (1) the relationship between the adoption of ITEM and B2BEC benefits; and (2) the impact of organizational characteristics (e.g. organizational IT maturity and IT evaluation resources (ITER) allocation) on the relationship between the adoption of ITEM and B2BEC benefits. The cross-national survey results provide empirical evidence in support of our proposed model, and demonstrate that: (a) the level of organizational IT maturity has a direct and significant impact on the adoption of ITEM; (b) the adoption of ITEM has a positive relationship with the ITER allocation; and (c) the ITER allocation has a significant direct influence on B2BEC benefits.

Keywords: e-commerce, IT evaluation, organizational characteristics, IT maturity.

1 Introduction

Large organizations are investing a significant amount of money in IT and have become increasingly dependent on it. According to International Data Corp. (IDC), global IT spending is forecast for compound annual growth of 6.3 to reach US$1.48 trillion in 2010 [22]. A 2005 survey by Gartner showed that on average organizations spent 6.1% of their gross revenue on IT [15] and the growth in IT spending as a percent of revenue is increasing faster than corporate sales [10].

However, many senior executives have discovered that they have not yet fully reaped the B2BEC benefits from their IT investment despite large investments in IT over many years [29]. Indeed, the realization of benefits from IT investment has been the subject of considerable debate by many academics and practitioners over the last few decades [5]. Some researchers (e.g. [20]) take the position that the confusion over the realization of B2BEC benefits is due to, among other things, lack of use of IT evaluation methodology to ensure successful evaluation of IT.

G. Psaila and R. Wagner (Eds.): EC-Web 2007, LNCS 4655, pp. 149–158, 2007.

Despite the focus by recent literature of the role played by IT evaluation on realization of B2BEC benefits, the relationships between organizational characteristics (e.g. organizational IT maturity and IT evaluation resources (ITER) allocation), IT evaluation methodology (ITEM) adoption, and B2BEC benefits remain unclear [6]. For example, Hackbarth and Kettinger [18] have found that there was some positive relationship between organizational IT maturity and the IT evaluation. Melville et al. [29] have pointed out that successful ITEM adoption depends heavily on right level of ITER allocation. The lack of more in-depth studies in this area has prompted the authors to develop a model of IT evaluation management to examine the linkages between these organizational characteristics and ITEM adoption and their impact on the realization of B2BEC benefits via a cross-national survey.

2   Theoretical Model

2.1   Proposed Model

Our research model for this study is derived from these two different perspectives – production economics and organizational perspectives. The production economics perspective establishes a direct connection between IT evaluation practices and B2BEC benefits. The organizational perspective, on the other hand, forms the basis for conceptualizing the linkage between technology adoption and diffusion and B2BEC benefits realized.

Production Economics View of IT Evaluation Management Process. The production economics perspective views organizational activities as economic production processes [4]. In these processes, input factors such as ITEM adoption can assist in transforming IT investments into outputs such as B2BEC benefits. Previous studies have shown that organizations that adopt ITEM can achieve better business performance or benefits (e.g. [29, 36]). For example, the adoption of ITEM serves as an essential learning tool for organizations and it assists in gathering support from key stakeholders [28]. It also helps to underscore best practices which organizations can apply lessons learnt to future IT investment processes and this has a significant impact on organizations' benefits from IT investments [29]. This is important as the use of ITEM assists organizations to focus on all aspects of an IT investment at regular intervals in order to evaluate realized benefits against their original objectives and to initiate corrective action [36].

However, the adoption of ITEM is a complex process for most organizations [26]. Many studies have revealed that many organizations fail to adopt ITEM [34, 35]. For example, it has been reported that that almost half of large Australian organizations [24] and Taiwanese SMEs [27] failed to adopt ITEM. This indicates that organizations need to consider not just the economic perspective but also the organizational perspective when investing in IT.

Organizational View of IT Evaluation Management Process. While the production economics perspective is adapted to examine the effect of ITEM on B2BEC benefits, the organizational perspective is used to view the contribution of the organizational characteristics in eliminating or minimizing the barriers of adopting ITEM. The technology adoption and diffusion theory stipulates that various kinds of barriers will eliminate or minimize the benefits of IT investment. The Limits-to-Value model proposed by Chircu and Kauffman [8] stated that there are many adoption barriers of IT investment which can prevent organizations in realizing benefits from their IT investment. The technology adoption and diffusion theory highlights the need for organizations to eliminate or minimize the adoption barriers by, for instance, increasing the level of organizational IT maturity and by allocating the required ITERs. This will greatly assist organizations in successfully adopting the ITEM in order to realize expected B2BEC benefits.

2.2 Research Hypotheses

We have first examined the effect of ITEM adoption on B2BEC benefits from the production economics view. Then we have considered the relationship between organizational characteristics and ITEM adoption and their effects on B2BEC benefits from the organizational perspective. The technology adoption and diffusion theory stipulates that organizational characteristics are needed to eliminate as much adoption barriers as possible in order to successfully adopt ITEM which in turn will lead to realization of B2BEC benefits. We therefore propose five hypotheses in our research model (see Fig. 1). The following five hypotheses are discussed below.

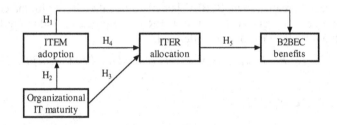

Fig. 1. A Proposed Model of IT Evaluation Management

IT Evaluation and B2BEC Benefits. According to the production economics perspective, ITEM adoption is viewed as input factor of economic production processes to assist in transforming IT investments into B2BEC benefits [4]. Previous studies have shown that organizations that adopt ITEM can achieve better business performance or benefits (e.g. [29, 36]). Thus, the following hypothesis is derived:

Hypothesis 1: ITEM adoption is positively related to B2BEC benefits.

Organizational Characteristics and IT Evaluation. The technology adoption and diffusion theory stipulates that various kinds of barriers will minimize the benefits of IT investment [8]. Therefore, organizational characteristics are needed in supporting the adoption of ITEM in order to realize B2BEC benefits. Two organizational characteristics are identified in this research – organizational IT maturity and ITER allocation. Organizational IT maturity refers an organization's existing IT infrastructure [3, 33] and its capability to utilize it to assist organizations in undertaking management processes [14] such as ITEM. Therefore, we argue that organizational IT maturity can influence the ability of organizations to adopt ITEM. Thus, the following hypothesis is derived:

Hypothesis 2: Organizational IT maturity is positively related to ITEM adoption.

The Limits-to-Value model proposed by Chircu and Kauffman [8] indicated that organizations need to eliminate or minimize the barriers for technology adoption in an orderly fashion. They argued that organizational barriers need to be eliminated or minimized before the resource barriers. Potential B2BEC benefits can only be realized when organizations take necessary steps to eliminate organizational barriers by, for example, increasing their IT maturity. Only then can organizations start to eliminate resources barriers by, for example, allocating the required ITER before they can convert potential B2BEC benefits into realized B2BEC benefits. We argue that ITEM plays a key role in the process as past studies have shown that organizations need to ensure that adequate ITERs are also allocated in the process of adopting ITEM [11, 31]. Thus, the following hypotheses are derived:

Hypothesis 3: Organizational IT maturity is positively related to ITER allocation.
Hypothesis 4: ITEM adoption is positively related to ITER allocation.

Organizational Characteristics and B2BEC Benefits. As noted earlier, organizations can expect to realize their B2BEC benefits once they have eliminated or minimized the resource barriers by allocating ITER in the process of adopting ITEM [8]. Therefore, the following hypothesis is derived:

Hypothesis 5: ITER allocation is positively related to B2BEC benefits.

3 Research Design

Questionnaires were sent to IT managers of large Australian and Taiwanese organizations. In total, 181_{Aus} and 217_{Twn} responses were received. In addition, late returns were compared with other response received earlier in order to check for non-response bias [2]. No significant differences were detected between the two samples on IT budget and total number of employees.

Table 1. Correlation Matrix for Constructs

	V1 IT maturity		V2 ITEM		V3 ITER		V4 B2BEC benefits		Alpha[a]		CR[b]	
	Aus	Twn	Aus	Twn	Aus	Twn	Aus	Twn	Aus	Twn	Aus	Twn
V1	.57[c]	.54							.84	.70	.84	.71
V2	.304[*]	.339[*]	.51	.76					.74	.90	.75	.86
V3	.136	.494[*]	.190[*]	.473[*]	.48	.64			.74	.82	.77	.78
V4	.347[*]	.282[*]	.187[*]	.204[*]	.482[*]	.378[*]	.60	.70	.80	.87	.81	.88

[*] p< .05 [a] Internal Consistency Reliability Cronbach's coefficient alpha. [b] Composite Reliability [13].
[c] The diagonal (in italics) shows the average variance extracted [13] for each construct.

Respondents were asked to indicate their agreement on a 6-point scale (1 for strongly disagree and 6 for strongly agree) with statements concerning four main constructs: (1) organizational IT maturity; (2) IT evaluation methodology adoption; (3) IT evaluation resources allocation; and (4) B2BEC benefits. The reliability analysis (Cronbach alpha) was conducted on these four main constructs. Cronbach's alphas and measurement for all constructs are all above 0.70 (see Table 1), indicating an acceptable reliability of the measures [30]. Table 1 presents the descriptive statistics and Pearson correlation for the constructs used in this survey. There was no evidence of multicollinearity between the independent variables as the coefficients were all under 0.50.

The theoretical model is operationalized as follows. The *organizational IT maturity* construct was based on Galliers and Sutherland's Revised Stages of Growth Model [14]. The revised model can be represented as six stages, each with its particular set of conditions associated with the four chosen elements: strategy, style, skills, and overall goals. These elements have provided a rich set of conditions upon which an organization can analyze and measure its organizational IT maturity.

The *IT evaluation methodology* construct was derived from Ward et al. [37]. The construct measured the levels of IT investment evaluation and benefit realization methodologies usage by organizations. This construct had three items and it measured the use of an IT investment evaluation methodology, the use of an IT benefits realization methodology, and the inclusion of intangible benefits in the IT evaluation process.

The *IT evaluation resources allocation* construct was derived from Chircu and Kauffman [8], Lin et al. [27] and Premkumar and King [32]. The other items were created by the authors. The construct had four items and measured the allocation of adequate technical skills, managerial skills, external vendor/consultant support, and other IT resources to undertake IT evaluation.

The *B2BEC benefits* construct was derived from Gregor et al. [16] and Lin et al. [27]. The construct measured the B2BEC benefits obtained in relation to the IT investments. The construct had four items. This included better relationships with suppliers/customers, improved quality of systems, greater efficiency, and employee satisfaction. In the absence of objective data on B2BEC benefits, the IT executives' perceptions were used. Although there has been some debate regarding the legitimacy of perceptual measures as a proxy for objective measures of B2BEC benefits, research has succeeded in alleviating some of the concerns by showing that perceptual measures of organizational performance has a strong positive relationship with more traditional objective measures [36]. Some studies have showed that there was a high degree of correlation between perceptual and objective performance measures in the process of measuring performance of several competing organizations [17, 36]. This reflects the degree of experience of IT executives and their constant involvement in the IT investment process. Finally, firm size was assigned as the *control variable*. Firm size was measured by the number of employees.

4   Data Analysis and Results

Most of the respondents' rank was above IT managers ($72\%_{Aus}$ and $69\%_{Twn}$). Most respondents were from manufacturing ($26\%_{Aus}$ and $56\%_{Twn}$), service ($37\%_{Aus}$ and

26%$_{Twn}$), and IT (7%$_{Aus}$ and 13%$_{Twn}$) industries. These respondents were typically large in revenue and number of employees by Australian and Taiwanese standards.

Prior to analysis, data were screened for possible outliers, and missing or out-of-range values. Next, missing values were estimated with the EM-algorithm in the missing value analysis (MVA) module in SPSS 11 (p-value = 0.024$_{Aus}$ and 0.653$_{Twn}$) and this is above the suggested value of p>0.000 [1]. All measures were then analyzed for reliability and validity in accordance with the guidelines set out by Jöreskog and Sörbom [22]. Multi-sample Confirmatory Factor Analysis (CFA) was used to construct a measurement model composed of three antecedent constructs of B2BEC benefits (IT maturity, ITEM adoption, and ITER allocation) and B2BEC benefits using maximum likelihood in LISREL 8.72. All constructs within the model were regarded as separate reflective measures. Overall, the resulting fit indexes indicated that the global measurement model fitted the data well: X^2 = 124.635 (112 degrees of freedom (df)), p-value = 0.195, comparative fit index (CFI) = 0.997, root mean square error of approximation (RMSEA) = 0.024, normed fit index (NFI) = 0.976, and expected cross-validation index (ECVI) = 0.812 is accepted between 90 percent confidence interval for ECVI = (0.780; 0.891). In addition, the group goodness-of-fit index (GFI) = 0.944$_{Aus}$ and 0.968$_{Twn}$ are also well above the acceptable level. Moreover, the ranges of all factor loadings and the measurement errors were acceptable and significant at alpha = 0.05, which provided evidence of convergent validity.

Furthermore, three types of validity were assessed to validate our measurement model: content validity, convergent validity, and discriminant validity. Content validity was established by ensuring consistency between the measurement items and the extant literature. This was done by interviewing IT executives and pilot-testing the questionnaire instrument. Churchill [9] has suggested that convergent and discriminant validities should be examined for construct validity. Therefore, we assessed convergent validity by examining composite reliability (CR) and average variance extracted (AVE) from the measures [19]. Our CR values of the four antecedent constructs were between 0.75~0.84$_{Aus}$ and 0.71~0.88$_{Twn}$ and all are above the suggested minimum of 0.70 (Hair et al., 1998). Their AVE values were between 0.48~0.60$_{Aus}$ and 0.54~0.76$_{Twn}$ and these values provided further evidence of convergent validity [13] (see Table 1). These AVE values could also be used to assess discriminant validity [13] which was evident in the results of this study as AVE values for all constructs were higher than the largest squared pairwise correlation between each construct (0.23$_{Aus}$ and 0.24$_{Twn}$) [12].

Table 2 shows the results of the multi-sample Structural Equation Modeling (SEM) for both the independent and dependent constructs. The overall fit of the proposed model was satisfactory (χ^2 = 180.303, df = 133, p = 0.004, RMSEA = 0.042, ECVI = 0.846, NFI = 0.964, and CFI = 0.989). Furthermore, the group goodness-of-fit indexes (GFI = 0.938$_{Aus}$ and 0.940$_{Twn}$) are also above the acceptable figure of 0.900 [19]. This shows that our IT evaluation management model is well suited for the multi-sample (i.e. Australia vs Taiwan) SEM analysis.

As indicated in Table 2, organizational IT maturity was positively related to ITEM adoption (β= 0.262***$_{Aus}$ and 0.310***$_{Twn}$), but had no significant direct impact on ITER allocation for the Australian sample (β= 0.064$_{Aus}$ and 0.307***$_{Twn}$). Therefore, **H2** is fully supported whereas **H3** is partially supported (supported in the Taiwanese sample but not in the Australian sample). Moreover, ITEM adoption was positively related to ITER allocation ($\beta= 0.101^{*}{}_{Aus}$ and $0.480^{***}{}_{Twn}$), but had no significant impact on B2BEC benefits ($\beta= 0.125_{Aus}$ and 0.053_{Twn}). Therefore, **H4** is fully supported while **H1** is not supported. **H5** is also fully supported as the ITER allocation had assisted both Australian and Taiwanese organizations in realizing their B2BEC benefits ($\beta= 0.568^{***}{}_{Aus}$ and $0.667^{***}{}_{Twn}$). Finally, the difference of beta (and gamma) coefficients of the two samples were tested and we found the relationships between IT maturity and ITER (**H3**) and between ITEM and ITER (**H4**) show that the Australian coefficients are significantly smaller than Taiwanese. In other words, country difference has an interaction effect on both **H3** and **H4** [23].

Table 2. The Results of Multi-sample Structural Equation Modeling (SEM)

Independent constructs	Dependent constructs					
	ITEM Beta (S.E.)[a]	Z test[b]	ITER Beta (S.E.)	Z test	B2BEC benefits Beta (S.E.)	Z test
IT maturity: AUS	0.262^{***} (0.082)	n.s.	0.064 (0.054)	-------		
TWN	0.310^{***} (0.082)		0.307^{***} (0.073)			
ITEM: AUS			0.101^{*} (0.045)	A < T	0.125 (0.107)	n.s.
TWN			0.480^{***} (0.093)		0.053 (0.070)	
ITER: AUS					0.568^{***} (0.110)	n.s.
TWN					0.667^{***} (0.112)	
R^2: AUS	0.06		0.04		0.29	
TWN	0.08		0.48		0.49	
Group model fit: AUS	GFI = 0.938					
TWN	GFI = 0.940					
Global model fit:	$\chi^2 = 180.303$		$d.f. = 133$		$p = 0.004$	RMSEA = 0.042
	ECVI = 0.846		NFI = 0.964		CFI = 0.989	

* p< .05; ** p< .01; *** p< .001   [a] All beta (and gamma) coefficients are unstandardized.
[b] The two beta coefficients between Australian and Taiwanese groups are estimated with approximate 95 percent confidence intervals. For example, if the intervals of Australian beta coefficient is smaller and not overlapping to Taiwanese, and then we declare A < T [22].

5 Discussions and Implications

To examine the relationship between the organizational characteristics and ITEM adoption as well as their effects on B2BEC benefits, this study developed a model of IT evaluation management with ITEM adoption and ITER allocation as the intermediate constructs linking organizational IT maturity to B2BEC benefits. The multi-sample SEM results from both Australian and Taiwanese organizations have revealed the support for the IT evaluation management model constructed earlier in the paper. Organizational IT maturity was found to have a significant positive impact on the ITEM adoption while the allocation of adequate resources to assist in adopting ITEM had resulted in realization of B2BEC benefits. These results have provided a number of insights and implications.

First, the results show that those organizations which had adopted ITEM alone could not produce B2BEC benefits. Second, organizations need to allocate adequate resources to successfully implement ITEM. It was found that there was a close linkage between organizational IT maturity and IT evaluation practices and, such linkage had assisted large organizations in realizing B2BEC benefits. Organizational IT maturity played a significant role in dictating IT evaluation methodology adoption which in turn led to allocation of adequate resources to undertake such a methodology. Third, as argued by Carr [7], successful IT investment comes down to successful management processes and practices, not necessarily those who invest in IT. This is a critical suggestion for those large organizations which need to assess their levels of organizational IT maturity carefully and consider the deployment of IT evaluation practices.

Moreover, this study makes a comparison between Australian and Taiwanese organizations in terms of their IT evaluation management process. Results from both Australian and Taiwanese organizations have largely supported the IT evaluation management model constructed earlier in the paper. However, some minor differences exist. The multi-sample SEM results indicated that organizational IT maturity and ITEM adoption had a more significant impact on ITER allocation within the Taiwanese sample. On possible explanation is that the mean of ITER allocation (M = 3.75) in the Taiwanese sample is significantly lower than the mean of ITER allocation (M = 4.32) within the Australian sample (F = 36.58, p= 0.00). This may indicate that organizational IT maturity and ITEM adoption are more imperative for Taiwanese organizations than for Australian organizations to increase their levels of ITER allocation (R^2 = 0.04_{Aus} and 0.48_{Twn}, see Table 2). Another possible explanation is that the average size of Taiwanese responding organizations (M = 2.87) is significantly smaller than the average size of Australian organizations (M = 4.29) (F = 32.21, p < 0.01). It has been found that in general smaller organizations which require more IT evaluation resources need to increase their organization IT maturity and this, in turn, will result in less barriers for adopting ITEM [25].

6 Conclusions

The purpose of the study is to examine the effects of the organizational characteristics and ITEM adoption on B2BEC benefits in Australian and Taiwanese large organizations. The empirical results have supported the model of IT management evaluation constructed. That is, organizational IT maturity significantly influenced ITEM adoption, and the realization of B2BEC benefits was positively affected by the allocation of adequate resources to undertake IT evaluation. This also implies that organizations interested in investing in IT should start with a thorough analysis of their organization in terms of their IT maturity. In this way, they can build realistic levels of organizational IT maturity in order to better equip themselves in adopting appropriate IT management processes such as IT evaluation methodology.

This study also draws attention to the fallacy of the adoption of ITEM alone will lead to successful outcomes for IT investment. Organizations that continue to spend significant financial resources on IT are often baffled by the return on their IT investment [28, 36]. Results from this study have suggested that large organizations should not only focus on their ITEM adoption and their levels of organizational IT maturity, but also should be more concerned with the increased allocation of adequate resources to undertake IT evaluation, since this is where B2BEC benefits can be directly monitored, managed, tracked, and realized.

Furthermore, some limitations in this research have to be acknowledged in this study. While there is much research on IT evaluation, there has been no research on the relationships between organizational IT maturity, ITEM adoption, ITER allocation, and the realization of B2BEC benefits. As a result, comparisons are difficult to make. In addition, this research has relied on the information provided at a particular point in time. Further research could take a longitudinal approach as the perception and management of IT evaluation practices are likely to change over time.

References

1. Arbuckle, J.L.: Full Information Estimation in the Presence of Incomplete Data. In: Marcoulides, G.A., Schumacker, R.E. (eds.) Advanced Structural Equation Modeling: Issues and Techniques, pp. 243–277. Lawrence Erlbaum, Mahwah New Jersey (1996)
2. Armstrong, J.S., Overton, T.S.: Estimating Nonresponse Bias in Mail Surveys. Journal of Marketing Research 14, 396–402 (1977)
3. Auer, T., Reponen, T.: Information System Strategy Formation Embedded into a Continuous Organizational Learning Process. Information Resource Management Journal 10, 32–43 (1997)
4. Banker, R.D., Davis, G.B., Slaughter, S.A.: Software Development Practices, Software Complexity, and Software Maintenance Performance: A Field Study. Management Science 44, 433–450 (1998)
5. Brynjolfsson, E., Hitt, L.M.: Computing Productivity: Firm-Level Evidence. The Review of Economics and Statistics 85, 793–808 (2003)
6. Byrd, T.A., Lewis, B.R., Bryan, R.W.: The Leveraging Influence of Strategic Alignment on IT Investment: An Empirical Examination. Information and Management 43, 308–321 (2006)
7. Carr, N.G.: IT Doesn't Matter. Harvard Business Review 8, 4–50 (2003)
8. Chircu, A.M., Kauffman, R.J.: Limits to Value in Electronic Commerce-related IT Investments. Journal of Management Information Systems 17, 59–80 (2000)
9. Churchill, G.A.: A Paradigm for Developing Better Measures of Marketing Constructs. Journal of Marketing Research 16, 64–73 (1979)
10. Computer Economics: IT Spending, Staffing, and Technology Trends 2006/2007 Study. Computer Economics Inc. (2006)
11. Eilat, H., Golany, B., Shtub, A.: Constructing and Evaluating Balanced Portfolios of R&D Projects with Interactions: A DEA Based Methodology. European Journal of Operational Research 172, 1018–1039 (2006)
12. Espinoza, M.M.: Assessing the Cross-Cultural Applicability of a Service Quality Measure: A Comparative Study between Quebec and Peru. International Journal of Service Industry Management 10, 449–468 (1999)
13. Fornell, C., Larcker, D.F.: Evaluating Structural Equation Models with Unobservable Variables and Measurement Error. Journal of Marketing Research 18, 39–50 (1981)
14. Galliers, R.D., Sutherland, A.R.: Information Systems Management and Strategy Formulation: 'The Stages of Growth' Model Revisited. Journal of Information Systems 1, 89–114 (1991)
15. Gomolski, B.: IT Spending and IT Adoption Profile, Gartner Research, ID Number: G00136380, pp. 1–11 (February 28, 2006)
16. Gregor, S., Fernandez, W., Holtham, D., Martin, M., Stern, S., Vitale, M., Pratt, G.: Achieving Value from ICT: Key Management Strategies. Department of Communications, Information Technology and the Arts, ICT Research Study Canberra (2004)
17. Grover, V., Teng, J., Segar, A.H., Fiedler, K.: The Influence of Information Technology Diffusion and Business Process Change on Perceived Productivity: The IS Executive's Perspective. Information and Management 34, 141–159 (1998)

18. Hackbarth, G., Kettinger, W.J.: Strategic Aspirations for Net-enabled Business. European Journal of Information Systems 13, 273–285 (2004)
19. Hair, J.F., Anderson, R.E., Tatham, R.L., Black, W.C.: Multivariate Data Analysis, 5th edn. Prentice Hall, Englewood Cliffs (1998)
20. Hitt, L.M., Brynjolfsson, E.: Productivity, Business Profitability, and Consumer Surplus: Three Different Measures of Information Technology Value. MIS Quarterly 20, 121–142 (1996)
21. IDC: Worldwide Spending on Information Technology Will Reach $1.48 Trillion by 2010, IDC Says. IDC Press Release (January 10, 2007)
22. Jöreskog, K.G., Sörbom, D.: LISREL 8: Structural Equation Modeling with the SIMPLIS Command Language. Scientific Software International, Chicago (1993)
23. Kline, R.B.: Principles and Practice of Structural Equation Modeling, 2nd edn. Guilford, New York (2006)
24. Lin, C., Pervan, G.: The Practice of IS/IT Benefits Management in Large Australian Organizations. Information and Management 41, 13–24 (2003)
25. Lin, C., Huang, Y., Tseng, S.: A Study of Planning and Implementation Stages in Electronic Commerce Adoption and Evaluation: The Case of Australian SMEs. Contemporary Management Research 3, 83–100 (2007)
26. Lin, C., Lin, K., Huang, Y., Kuo, W.: Evaluation of Electronic Customer Relationship Management: The Critical Success Factors. The Business Review, Cambridge 6, 206–212 (2006)
27. Lin, K., Lin, C., Tsao, H.: IS/IT Investment Evaluation and Benefit Realization Practices in Taiwanese SMEs. Journal of Information Science and Technology 2, 44–71 (2005)
28. Love, P.E.D., Irani, Z., Standing, C., Lin, C., Burn, J.: The Enigma of Evaluation: Benefits, Costs and Risks of IT in Small-Medium Sized Enterprises. Information and Management 42, 947–964 (2005)
29. Melville, N., Kraemer, K., Gurbaxani, V.: Review: Information Technology and Organizational Performance: An integrative Model of IT Business Value. MIS Quarterly 28, 283–322 (2004)
30. Nunnally, J.C.: Psychometric Theory. McGraw-Hill, New York (1978)
31. Ooi, G., Soh, C.: Developing an Activity-based Costing Approach for System Development and Implementation. Database for Advances in Information Systems 34, 54–71 (2003)
32. Premkumar, G., King, W.R.: Organizational Characteristics and Information Systems Planning. Information Systems Research 5, 75–109 (1994)
33. Schuh, R.G., Leviton, L.C.: A Framework to Assess the Development and Capacity of Non-profit Agencies. Evaluation and Program Planning 29, 171–179 (2006)
34. Standing, C., Guilfoyle, A., Lin, C., Love, P.E.D.: The Attribution of Success and Failure in IT Projects. Industrial Management and Data Systems 106, 1148–1165 (2006)
35. Standing, C., Lin, C.: Organizational Evaluation of the Benefits, Constraints and Satisfaction with Business-To-Business Electronic Commerce. International Journal of Electronic Commerce 11, 107–153 (2007)
36. Tallon, P.P., Kraemer, K.L., Gurbaxani, V.: Executives Perceptions of the Business Value of Information Technology: A Process-Oriented Approach. Journal of Management Information Systems 16, 145–173 (2000)
37. Ward, J., Taylor, P., Bond, P.: Evaluation and Realization of IT Benefits: An Empirical Study of Current Practice. European Journal of Information Systems 4, 214–225 (1996)

Linking M-Business to Organizational Behavior Levels – A Mobile Workforce Centered Research Framework

Daniel Simonovich

European School of Business,
Reutlingen University, Alteburgstrasse 150, D-72762 Reutlingen
daniel.simonovich@reutlingen-university.de

Abstract. The potential of mobile electronic business (m-business) applications and wireless technologies has created interdisciplinary interest among researchers in the fields of computer science, business administration, and the social sciences. However, one preeminent inter-discipline has yet found little systematic inclusion: organizational behavior (OB), a study field of business administration relating to psychology. This paper highlights the three levels of OB and articulates typical questions with behavioral emphasis arising from the use of mobile applications in organizations. Empirical insights on the mobile workers gained across nine industries suggest that any research agenda addressing the effect of m-business on OB levels should reflect industrial differences. Building on the mobile workforce concept, a three-dimensional reference framework is presented as a guide for a behavioristic research agenda.

Keywords: Organizational behavior, work relationships, m-business.

1 Introduction

Mobile business (m-business) applications have received considerable and ongoing attention over the last few years [1] [2]. Though mainly propelled by technological advances, research into the application of mobile technologies is far from being limited to the computer science discipline. While information and communication technology (ICT) has certainly played an enabling role by advancing mobile communication protocols [3], quality of service [4], systems integration [5] and security aspects [6], business studies and social sciences have joined the research efforts in many ways (see fig. 1). Both business administration and the social sciences have produced research outcomes relating to m-business within their respective research traditions. For example, human mobility has been a long standing study subject in sociology while many business studies were dedicated to estimate the market success of various mobile applications [7]. In addition, important m-business related research questions have been explored at the boundaries between business, social sciences, and computer science. One example is the analysis of user acceptance as a prerequisite for m-business market success [8].

Despite such interdisciplinary research interest, an inter-discipline at the boundary has hardly been systematically explored: organizational behavior (OB) – a managerial

G. Psaila and R. Wagner (Eds.): EC-Web 2007, LNCS 4655, pp. 159–168, 2007.
© Springer-Verlag Berlin Heidelberg 2007

study field dedicated to understanding human psychology in organizations [9]. However, since mobile applications impact the organizations they penetrate, a profound effort for understanding the effects of mobile applications on behavioristic leadership issues in these organizations seems worthwhile. For example, novel information and communications means of mobile applications may increase the amount of central control and power over parts of the mobile workforce and thereby alter an organization's culture. Consequently, contemporary behavioral psychology analyses would be incomplete without the appreciation of modern applications support among the mobile workforce.

The next chapter introduces the three manifested levels of organizational behavior analysis: Individual, group, and organizational [10]. It sketches organizational psychology questions in regard of the m-business phenomenon. Chapter 3 recaptures the spectrum of mobile applications from a corporate business perspective. This is to relate m-business to OB at an appropriate level of analysis. Chapter 4 links mobile applications to OB via the mobile workforce notion for the purpose of framing a research framework. It builds on an empirical pre-study into the mobile workforce to do so. The concluding chapter outlines future research opportunity.

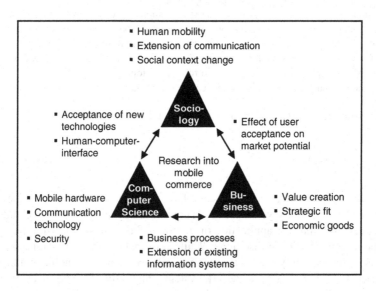

Fig. 1. Overview of multi-disciplinary research into the application of mobile technology

2   Analysis Levels of Organizational Behavior (OB)

Organizational behavior (OB) has become the academic subject responsible for collecting knowledge about the nature and psychology of human beings in the context of an organization. It is not an original discipline on its own but a collection of loosely related concepts from the disciplines of sociology, psychology, anthropology, political science, economics, industrial engineering, and medicine [9]. Nine out of ten international postgraduate management programs require OB as part of their core curriculum [10]. OB addresses behavior at the individual, group and organizational levels and is also concerned with the relationship between the individual and the group, and how both interact with the organization [11]. From a practical point of view, the purpose of OB is the application of the collective wisdom of human behavior in organizational decision-making with an emphasis on leadership aspects [12]. Before turning to m-business and its potential significance for OB related issues, it is useful to distinguish two quite opposite motivations in interdisciplinary m-business research (fig 2).

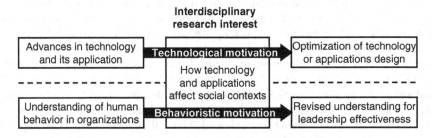

Fig. 2. Technological versus organizational behavior motivation in interdisciplinary research

The technologically driven motivation seeks to understand technology in social contexts [13] in order to optimize technical features. The design of human-computer-interfaces in mobile devices [14] and the development of context aware applications [15] are examples thereof. By contrast, behavioristic research initiatives look at technology to revise or upgrade the understanding of people in organizations as well as potential implications on organizational leadership issues. Research initiatives like the recently established Workshop on Mobile and Networking Technologies for Social Applications (MONET) can be seen as a contribution in that direction [16]. The analysis of group dynamics in regard of Bluetooth enabled GUIs during a scientific conference is an example of a behavioristic interest in social contexts [17]. Similar work exists about university student and staff interaction [18]. However, these recent contributions do not exactly position as OB, a discipline concerned with industrial organizations (rather than academic communities) and corresponding leadership issues across three distinct but interrelated levels of analysis (fig. 3).

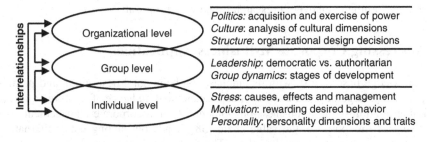

Fig. 3. Overview of typical organizational behavior issues across three analysis levels

It is worthwhile reviewing those analysis levels while highlighting exemplified issues within theses levels. The individual level is dedicated to understanding human diversity along complex topics such as personality [19], motivational attitude [20], or stress behavior [21]. Various questions might be raised in regard of discontinuous technology and applications developments such as mobile applications: How do differences in personality [22] influence the adoption and effective use of these novel applications? In how far do they affect motivational categories [23]? To what extend do m-business applications give emphasis to stressors [24] and their effect on stress symptoms? At the intermediate analysis level OB is known for themes such as group dynamics [25] or leadership effectiveness [26]. These subjects might also deserve study in regard of m-business developments: In how far may intra- and inter-organizational m-business applications accelerate or slow down the stages of collaborative group development [27] known as forming, storming, norming, and performing? How effectively do alternative leadership styles [28] support the collaboration dynamics of corporate m-business applications? Finally, we can look at the third and probably most complex analysis layer. Besides the formal topic of organizational design [29], the organizational level includes hard-to-grip concepts such as corporate culture [30] or power and politics [31]. Challenging questions arising here are, for example: To what extend do m-business applications support the transformation from traditional organization concepts such as functional, divisional, or matrix design [31] to modern organization principles such as process based or virtual organizations [32]? In how far do some unwritten organizational rules change and thereby alter the cultural parameters of an organization [33]? Does the pervasiveness of m-business applications increase central control or modify the relative influence and power [34] of roles within an organization?

The questions above are just examples. However, they illustrate that OB-driven inquiry tends to be way more behavioristic than the majority of interdisciplinary m-business research contributions – after all the results of the latter tend to be predominantly published in computer science or MIS related journals and conferences. The observation holds that despite some historical tradition in the social sciences, mobility has not quite received the attention of leading OB journals throughout this decade [35] [36] [37] (the period when research into mobile business has become weighty and evidence-based). Before pushing the focus of interdisciplinary m-business examination more towards the behavioristic OB edge, it is nonetheless important to find some effective and straightforward way for relating the universe of mobile applications to the kind of questions above. The next chapter deals with this task before turning to an overall research framework proposal.

3   Addressing Corporate M-Business Applications

The knowledge base of OB, with its emphasis on behavioral description and prediction, is by and large targeted at an applied management audience. Therefore, it seems advisable to address the m-business phenomenon from a corporate management perspective. For this purpose, the value chain [38] stands out among many alternative ways of structuring an application class. The value chain structure has successfully been deployed to communicate the business meaning of traditional management

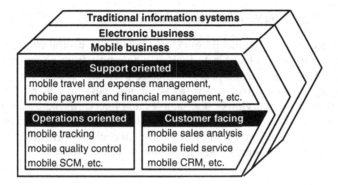

Fig. 4. Value chain oriented structure for analyzing m-business applications from the organizational perspective

information systems (MIS) [39], e-business [40], and m-business [41] to a managerial audience. Fig. 4 gives a simplified overview of this approach.

The value chain strives to give an overall picture of a firm's activities. These activities may be logically grouped into operations oriented, customer facing, and support oriented occupations. In traditional MIS, these categories loosely relate to the software families of SCM, CRM, and ERP and can be extended to m-business applications (see fig. 4). The convenience of this general-purpose framework, however, comes at a cost: Differences across industries are rather neglected such that cross industry applications are given preference over industry specific applications. As a consequence, administrative users ("white collar workers") may be overrepresented compared to industrial or technical application users ("blue collar workers") of mobile applications. Before crafting a framework for guiding OB oriented m-business research, we shall look at the results of a survey which provides insight into the extent of industry specificity of the mobile work force.

4   Proposal of an OB Oriented Research Framework

The notion of mobile workforce, the subset of employees with a significant portion of physical movement, offers a natural link between m-business applications and an organizational behavior (OB) perspective. On one hand, any interest in human behavior in the context of mobility strictly references the mobile workforce. On the other hand, the users of m-business applications can reasonably be assumed to move geographically. Over the last 10 years, a number of studies have dealt with better understanding the mobile workforce in regard of emerging m-business applications. The Mobile Informatics Research Program at the Swedish Information Technology Research Institute (SITI) is a prominent example which produced a number of studies including mobile field work, mobile IT for public safety, or mobile sport informatics [42]. These studies capture important lessons learned about the way mobile workers interact. However, to address mobile business for the purpose of OB oriented examination, a more global investigation into the mobile workforce – at the inclusion of core industries – is desirable. The following section describes the results of such empirical investigation undertaken.

4.1 Empirical Investigation into the Mobile Workforce

For the purpose of using the mobile workforce notion as a neutral concept connecting m-business to OB, it is advisable to look at it foremost from an applications independent perspective. For this purpose, the mobile workforce of nine core industry was analyzed using a quantitative survey (n=540 companies) as well as the qualitative results of 30 in-depth interviews – the samples and interviews were gathered in German speaking countries. Fig. 5 depicts the mobile workforce as percent of total staff. It also introduces the dimension of "blue collar" (technically oriented) versus "white collar" (administration oriented) employees among mobile workers.

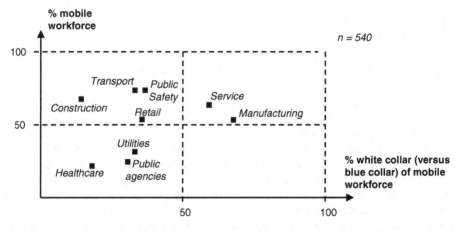

Fig. 5. Industry specific mobile workforce percentage and white versus blue collar distribution

The relative distribution between blue collar and white collar may be seen as an indicator of industry specific (technical) versus cross industry (general administration) potential for corporate mobile applications support. Three industry analysis clusters emerge by looking at the quadrants: Mobile blue collar dominant industries, non mobile blue collar dominant businesses, and white collar dominant enterprises. This underlines the existence of generic differences in mobile nature across industries. Another applications-independent aspect of the mobile workforce is an industry's dominant interaction topology in terms of communication with a central site versus peer-to-peer communication among mobile workers (see figure 6). With respect to organizational behavior, this distinction is particularly relevant for investigation at the group and organizational levels.

Besides, the empirical results also confirmed considerable industry differences concerning the portion of external mobile interaction (with outside parties) versus internal – with an external range from 15% for public safety to 75% for healthcare. Overall, these findings suggest that the study of mobility at the workplace should to some extent account for industry characteristics. Industry specificity is, of course, not a novel distinction in m-business, as witnessed by many m-business case studies and market potential forecasts. The data shown in figures 5 and 6, however, is different

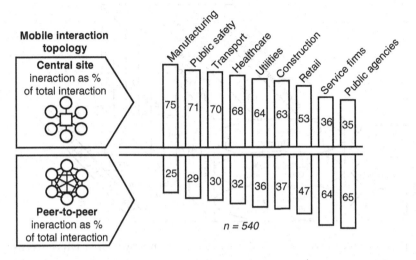

Fig. 6. Centralized versus peer-to-peer topology as % of total mobile interaction

from these studies in that it offers an aggregate and applications-independent mobility outlook of core industries. The following section synthesizes the introduced concepts of OB, m-business, and mobile workforce into an integrative research framework.

4.2 A Research Framework for Studying OB Issues in M-Business

From the discussion so far it should be clear that the study of m-business from an OB point of view is a complex task: First, OB contains many relevant but multifaceted issues along three analysis levels. Second, the mobile applications universe is wide and increasing. While a value chain or other general-purpose structure offer guidance through organizational classification, these simplifications do not typically account for industry specificity. The mobile workforce insights above, however, offer a first distinction of industries by their mobile workforce dominance and an indicative measure of industry specificity (blue collar percentage). Figure 7 brings together these considerations in a three-dimensional research framework. The levels of OB and contemporary issues within these levels are brought to the foreground while segmenting application clusters according to a value chain mindset. The third dimension is generated by the industry perspective.

The nine basic industries studied are clustered according to mobile workforce domination and white versus blue collar emphasis. Ultimately, this framework leads to a research array offering a focus of OB determined research inquiry for m-business application categories in specific industries. One of these array elements, for example, may be dedicated to study the cultural effects (organizational OB level) of mobile travel and expense applications (support oriented application) in service firms (white collar dominant). Without prejudging the specific choice and research methods, the

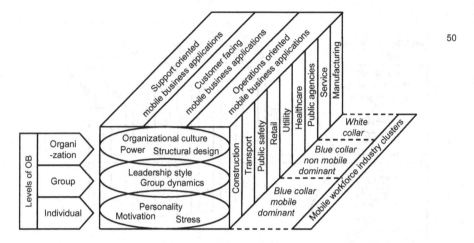

50

Fig. 7. Mobile workforce centered m-business research array from the OB perspective

framework of figure 7 could help facilitate the execution buildup of an m-business research agenda from an OB perspective.

5 Conclusions

Electronic and mobile business research is increasingly taking advantage of interdisciplinary synergy. However, at a closer glance, strictly behavioristic research inspirations seem underrepresented. This applies specifically to organizational behavior (OB), a central and practically relevant management studies field which examines business psychology at different levels of organizational granularity. So far, the leading OB forums have not included mobility or mobile applications among their contemporary research themes [35] [36] [37]. The notion of mobile workforce – those employees who demonstrate a significant degree of physical mobility – was introduced as a natural interface and neutral ground between the study of mobile applications on one hand and the study of human behavior in organizations on the other. Empirical insights at aggregate industry levels were used to position industries according to the degree and nature of their respective mobile workforce. The resulting three-dimensional examination framework offers a behavioristic research agenda for cumulating and integrating m-business related OB research output into a coherent reference model. Three directions might characterize further research: First, the framework can be used to revisit some existing m-business work related to social contexts. Second, further behavioristic contributions might be directed to fill out the proposed examination framework. From a research methods point of view, it is noteworthy that behavioristic research is usually an empiricism-based theory building process whereas the computer science and MIS research communities find themselves predominantly in a design science research tradition [43]. Finally, the proposed research framework itself might be subject to evolution as a consequence of results of the first two further research directions or related work.

References

1. Hu, W.C., Lee, C.W., Yeh J.H.: Mobile Commerce Systems. In: Shi, N.S. (ed.): Mobile Commerce Applications. Idea Group Publishing, Hershey, pp. 1–23 (2004)
2. Fouskas, K.G., Giaglis, G.M., Kourouthanassis, P.E., Karnouskos, S., Pitsillides, A., Stylianou, M.: A roadmap for research in mobile business. International Journal of Mobile Communications 3(4), 350–373 (2005)
3. Chlamtac, I.: Editorial - Special Issue on Protocols and Software Paradigms of Mobile Networks. Mobile Networks and Applications 3, 121–122 (1998)
4. Markasin, A., Olariu, S., Todorova, P.: QoS Oriented Medium Access Control for All - IP/ATM Mobile Commerce Applications. In: Shi, N.S. (ed.) Mobile Commerce Applications, pp. 303–331. Idea Group Publishing, Hershey (2004)
5. Velasco, J., Castillo, S.: Mobile Agents for Web Service Composition. In: Bauknecht, K., Toja, A.M., Quirchmayr, G. (eds.) E-Commerce and Web Technologies. Proceedings of the 4th Internatinal Conference EC-Web, Springer, Heidelberg (2003)
6. Veijalainen, J., Visa, A.: Security in Mobile Computing Environments. Mobile Networks and Applications 8, 111–112 (2003)
7. Samuelsson, M., Dholakia, N.: Assessing the Market Potential of Network-Enabled 3G M-Business Services. In: Shi, N.S. (ed.) Wireless Communications and Mobile Commerce, pp. 23–48. IRM Press (2004)
8. Ogertschnig, M., van der Heijden, H.: A Short Form Measure of Attitude Towards Using a Mobile Information Service. In: Proceedings of the 17th Bled Electronic Commerce Conference, Bled (2004)
9. McKenna, E.: Business Psychology and Organizational Behavior. 3rd edn. Psychology Press, Hove, 4 (2000)
10. Navarro, P.: What the best MBAs know - how to apply the Greatest Ideas taught in the Best Business Schools, pp. 6–15. McGrawHill, New York (2005)
11. Griffin, R.W.: Organizational Behavior - Managing People and Organizations, pp. 1–26. Houghton Mifflin, New York (2006)
12. Wood, J.: Deep roots and far from a soft option. In: Dickson, T., Bickerstaffe, G. (eds.) Mastering Management, pp. 217–224. Pitman Publishing, London (1997)
13. Kurvinen, E., Oulasvirta, A.: Towards socially aware pervasive computing: A turntaking approach. In: Proceedings of the International Conference on Pervasive Computing and Communications (PerCom), Orlando, Florida, pp. 346–351. IEEE Computer Society, Los Alamitos (2004)
14. Lee, Y.E., Benbasat, I.: Interface design for mobile commerce. Communications of the ACM 46(12), 48–52 (2003)
15. Schilit, B.N., Adams, N.I., Want, R.: Context-aware computing applications. In: Proc. Workshop on Mobile Computing Systems and Applications, Santa Cruz, USA, pp. 85–90. IEEE Computer Society, Los Alamitos (1994)
16. Ferri, F., Rafanelli, M., D'Ulizia, A.: MONET 2006 Co-chairs. In: Meersman, R., Tari, Z., Herrero, P. (eds.) On the Move to Meaningful Internet Systems 2006. LNCS, vol. 4277, p. 873. Springer, Heidelberg (2006)
17. Nicolai, T., Yoneki, E., Behrens, N., Kenn, H.: Exploring Social Context with the Wireless Rope. In: Meersman, R., Tari, Z., Herrero, et al. (eds.) On the Move to Meaningful Internet Systems 2006. LNCS, vol. 4277, pp. 874–883. Springer, Heidelberg (2006)
18. Eagle, N., Pentland, A.: Reality mining: Sensing complex social systems. Personal and Ubiquitous Computing 10, 255–268 (2006)

19. Wright, D.S., Taylor, A., Davies, D.R., Sluckin, W., Lee, S.G.M., Reason, J.T.: Introducing psychology: an experimental approach. Penguin (1970)
20. Maslow, A.H.: Motivation and personality, 3rd edn. Harper & Row, New York (1954)
21. Briner, R., Reynolds, S.: Bad theory and bad practice in occupational stress. Occupational Psychologist 19, 8–13 (1993)
22. Hampson, S.: State of the art - Personality. The Psychologist, June, 284–288 (1999)
23. Alderfer, C.P.: Existence, relatedness, and growth: human needs in organisational settings. Free Press, New York (1972)
24. Murray-Bruce, D.J.: Promoting employee health. Macmillan, Basingstoke (1990)
25. Blau, G.: Influence of group lateness on individual lateness: A cross-level examination. Academy of Management Journal, 1483–1496 (1995)
26. Hersey, P., Blanchard, K.H., Johnson, D.E.: Management of Organizational Behavior - leading Human Resources, 8th edn., pp. 107–125. Prentice Hall, Upper Saddle River (2010)
27. Tuckmann, B.W.: Development sequence in small groups. Psychological Bulletin 63, 384–399 (1995)
28. Tannenbaum, R., Schmidt, W.H.: How to choose a leadership pattern. Harvard Business Review, 162–180 (1973)
29. Sadler, P.: Designing Organizations, pp. 16–21. Kogan Page, London (1998)
30. Chatman, J.A., Jehn, K.A.: Assessing the relationship between industry characteristics and organizational culture: How different can you be? Academy of Management Journal, 522–553 (June 1994)
31. Silbiger, S.A.: The ten day MBA, pp. 139–145. Quill, New York (1993)
32. Alexander, M.: Getting to grips with the virtual organization. Long Range Planning 30(1), 122–124 (1997)
33. Ferraro, G.P.: Cultural Dimensions of International Business, 3rd edn. Prentice Hall, Upper Saddle River (1997)
34. Burnham, D.H., McClelland, D.C.: Power is the great motivator. Harvard Business Review, 134 (1995)
35. Kramer, M.R., Staw, B. (eds.): Research in Organizational Behavior - an Annual Series of Analytical Essays and Critical Reviews (2000-2006)
36. Rosser, J.B. (eds.).: Journal of Economic Behavior and Organization (2000-2006)
37. Harrison, D.A. (ed.).: Organizational Behavior and Human Decision Processes (2000-2006)
38. Porter, M.E.: Competitive Advantage - Creating and Sustaining Superior Performance, pp. 33–61. Free Press, New York (1985)
39. Millar, V.E., Porter, M.E.: How Information gives you Competitive Advantage. Harvard Business Review, 149–160 (1985)
40. Porter, M.E.: Strategy and the Internet. Harvard Business Review, 63–78 (2001)
41. Simonovich, D.: Business-Centric Analysis of Corporate Mobile Applications. In: 8th International Workshop on Mobile Multimedia Communications (MoMuC), pp. 269–273 (2003)
42. Kristoffersen, S., Ljungberg, F.: Innovation of IT use in mobile settings. SIGCHI Bulletin 31(1), 29–34 (1999)
43. Hevner, A.R., March, S.T., Parc, J., Ram, S.: Design Science in IS Research. Information Systems Quarterly 28(1), 75–85 (2004)

Prediction of Keyword Auction Using Bayesian Network

Liwen Hou[1], Liping Wang[1], and Kang Li[2]

[1] 535, Fahuazhen Rd., Department of Management Information System
Shanghai Jiaotong University, Shanghai, China
[2] 415 Boyd GSRC, Department of Computer Science, University of Georgia,
Athens, Georgia, USA 30602

Abstract. Online keyword auctions, in which marketers bid for advertising slots along the search engine results, have become a new channel of advertisement. To better manage the advertisement campaign, a key challenge for advertisers is to predict each keyword's bidding price and effectiveness (e.g. click through rate), which are not priorly known to the individual advertiser. This paper identifies those relevant variables affecting auction strategy and models them in causal connections using history data in order to simulate the bidding behavior. We verified the effective necessaries of these predictions using empirical auction data, and our result indicated that the prediction with Bayesian Network produce close-to-reality results.

Keywords: Bayesian Network, Keyword Auction.

1 Introduction

The recent success of advertisement through Google and other search engines have re-ignited the online marketing by keywords auctions and indicated huge potentials. Keyword auctions have brought new opportunities to companies (especially smaller ones that can't afford traditional advertisement media, such as radio and TV) to reach a large but targeted audiences with a low marketing cost. However, as a new advertisement channel, marketers using online keyword auctions also face a serious challenge – how to optimally manage the auction strategy to maximize the budget.

Because cost and effectiveness (e.g. click through rate) of each keyword varies dynamically due to competition and user dynamics, it is extremely hard for advertiser to achieve a high payoff rate without consider them. For example, as the number of competitors of a keyword and their auction prices changes, a marketer might see his advertisement position drops. On one hand, increasing the auction budget on a particular keyword might get the link to his site in a better advertising position. On the other hand, if competition for this keyword is too intense, it might be good to shift budget to other keywords. When advertisement campaign is not well management, some of advertisers even claim mainstream auction platforms, like google and yahoo, reap much from the novel business model while auctioneers may suffer because of knowing nothing about the bidding strategy, just as what *Business Week* described "stupid money".

G. Psaila and R. Wagner (Eds.): EC-Web 2007, LNCS 4655, pp. 169–178, 2007.
© Springer-Verlag Berlin Heidelberg 2007

Research on this area tends to resort to the classical auction theory, which, however, does not match perfectly with the keyword auctions, especially for the fact that advertisers updating their auction bid instantly and continuously.

This paper takes a learning approach and presents a model consisting of a Bayesian network to predict the bidding results, which can help maximize benefit for a given budget constraint. The contributions of this paper are in two folds. First, to the best of our knowledge, the work is the first to apply Bayesian network to predict auction results. The prediction is around three key variables (*bid price*, *positions*, and *click through rate*) affecting auction strategy. Second, we verified the prediction results using empirical auction data obtained from the a highly competitive market (flower-retail) on the top two search engines, *Google* and *Baidu*, in China.

The rest of the paper is organized as follows: Section 2 reviews the related literature in the area of online auction prediction and management. Section 3 presents our model with Bayesian network predictions. Section 4 describes our evaluation of the auction prediction using empirical data. Finally, Section 5 concludes the paper.

2 Related Work

Research in the area of online keyword auction is just in the bud. Related literatures are mostly very recent, and they can be categorized two kinds with respect to the methodology. Here we only survey those closely related to the computer science area due to space limitation.

Bidding strategy is fundamental target of keyword auction, which incurs many kinds of interests. Research by Animesh et al (2005) took into account different profiles of bidding strategy and examined the reasons underlaid across customer's quality signaling. Besides, he also examined the relationship between advertisers' quality and their bidding strategies in online setting. Benjamin et al (2005) aimed at strategic behavior of keyword auction. They found strategic behavior impaired the benefit of both search engine and market. Another issue is mechanism adopted by various search engines. Juan et al (2005) inspected two kinds of mechanism with respect to the correlation of advertiser's willingness to pay and terms relevance and verified that one with the product of bid and click through not only is more robust than the other but also converges to optimal point in revenue of advertisers. Simulation as a most popular technology is widely adopted in this area. Brendan et al (2004) simulated PPC (pay per click) keyword auction based on agent technology. The simulation result is used by a profit function and succeeded in real bidding process on Overture platform. Our paper is much like this one except the simulation technology and parameters estimation. Agent technology is also used by Sunju et al (2004) who elaborately induced agents to smartly bid with a tractable result afterward based on Markov chain model as well as identified the optimal bidding strategies by experiment. Young-Hoon (2005) posited an integrated framework including four key components to simulate bidding behavior for keyword auctions on website and concluded that willingness to bid is the only pinpoint. Similar vehicles are agent and machine learning that are applied in the research of Maria-Florina (2005) and Fang-Rui et al (2004) although the latter appeared in other papers on electronic commerce and procurement rather than only on keyword auction. In addition other research like those done by Andreina and David et al (2005) also enriches the extant literature body. By this token online keyword auction as the novel area is catching a wide attention.

3 Bayesian Network Based Prediction Model

3.1 Methodology

Many factors pertain to keyword bidding process. In this paper we do not cover all of them but instead focus on three crucial ones: *bid, position* and *click through (abbr. CT)*. These three are essential components of bidding process and form a casual chain to be able to trace the whole campaign. In most of the search engine bidding process, the bidding price of all competitors determine the relative ads positions, with top position usually attract the most clicks. Thus *bid, position* and *CT* form a cause and consequence chain. Since an advertiser's consequence (position and CT) depends on both its bidding price as well as bidding from others and user clicks, the outcome are unknown by nature. To guide users, this paper uses Bayesian network to make the bidding result prediction.

Bayesian network is a popular tool to build a complicated structure and casual relationship formed in a context of various factors. For example, Sucheta (2004) illustrated how domain knowledge of expert can be extracted and nested into a casual map, which ultimately consists of a bayesian network. The popularity of bayesian network comes from the following properties: a) if multivariable model is taken into account their correlation, such as the regression, bayesian network can be effectively induced using this technique even in the absence of some data; b) bayesian technique can fuse prior knowledge and experiences of bidders into a probability model and reveal strengthen of variables in a causal context; c) bayesian network can effectively avoid over-fitting of data and obey the internal connection of factors.

Bayesian networks fits perfectly to the problem of online bidding because of the following factors: first many factors, not only bid and rank but behavioral profiles, influence the effectiveness as well; second, not all those factors directly control even decide the final bidding result. Actually some of factors are intermediate that transfers influence from others to the critical variables.

In the rest of the work, two bayesian network models (differentiated by whether considering the attraction of website per se) are developed for the sake of predicting each of the parameters of bidding and deducting a reasonable strategy by which advertiser can achieve the expected objectives in the next campaign. The models are examined using the history data from a flower vendor in China and proven effective.

3.2 Prediction Model Without Website Attraction (Scenario 1)

3.2.1 Model Structure

Bayesian network generally represents a suite of model that consists of two kinds of sub-model. One is a qualitative structure model and another is a quantitative parameter model. The former describes the casual connection of factors and the latter indicates the direction and magnitude of the influence among those factors. It is relatively easy to draw a Bayesian diagram in terms of casual chain, just as Figure 1. However, how to derive these conditional probabilities are still very challenging. This is also the knob of the model because that is the concerning of parameter model which

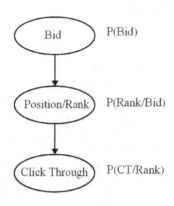

Fig. 1. Bayesian Model for Scenario 1

demonstrates the magnitude of influence of factors in the model. If the interaction is relatively weak the casual relationship linking two factors in the network diagram may be crossed off to show that there actually exist no casual connection to them. So one need to estimate a specific probability in this diagram by virtue of real data, which is called learning.

Figure.1 shows that *bid* is the reason of which *position/rank* is held and the latter is sequentially the reason of how many *CT* is allured. These connections consist of the structure model, which can be expressed as the following probability function:

$$P(Bid, Rank, CT) = P(Bid) * P(Rank \mid Bid) * P(CT \mid Rank) \tag{1}$$

The left of the formula (1) should also satisfy formula (2) according to conditional probability formula.

$$P(Bid, Rank, CT) = P(Bid) * P(Rank \mid Bid) * P(CT \mid Bid, Rank) \tag{2}$$

Combine formula (1) and (2) one can realize an implied condition is held:

$$P(CT \mid Rank) = P(CT \mid Bid, Rank) \tag{3}$$

This formula verifies a common fact that *CT* is independent to *bid* if *rank* is given.

Each probability, *P(bid)*, *P(Rank|Bid)* and *P(CT|Rank))*, corresponds to a specific casual relationship in Fig.1 and can be calculated from a dataset which is the true value. One can deduce the corresponding posterior probabilities from those prior probabilities when new data input. This learning process help develop a stable casual connection among those factors.

3.2.2 Parameters Learning

Parameter model is in effect the result of parameter learning conditioned on the structure model in the bayesian network. The learning process is a process to infer posterior probability from prior probability by continuous training using history data. In this paper *P(Rank|Bid)* and *P(CT|Rank)* are trained and adjusted using sample data of *bid* in order to successfully estimate a new situation for each click.

One necessary step to apply most of machine learning algorithms (such as Bayesian networks) is to discretize continuous variable. This includes two steps: first, the number of values for each continuous variable to be discretized should be fixed but sufficient to imitate the shape of the variable. Second, small intervals encompassing only one discretized value in each of them will be initiated to confine the scope of value of continuous variable.

In this paper two techniques, equal interval method (no monitoring) and weighted information-loss discretization (monitoring) (Jensen, 2001), are applied to achieve the discretization of *bid, position* and *CT*. *Position* per se is discrete while *bid* and *CT* are continuous so the both is discretized into 200 intervals and *position* is supposed to change within 30 (too a large value to be worthy of bidding). Each conditional probability for any parameter in Fig.1 is assumed to uniformly distribute before learning but will be substantially adjusted to acquire a more accurate parameter model by learning those history data. Maximal likelihood estimation (MLE) is applied to bear up the learning process with a refinement to the result in order to synthesize prior experience and knowledge of bidding. A function in BNT, a matlab toolbox developed by Kevin Murphy, learn_params_em is modified so that its four parameters below can reflect the keywords bidding situation:

- jtree_inf_engine: reasoning engine in the parameter learning;
- cases: dataset for learning;
- maxiter: the number of loops;
- epsilon: stop condition for loop.

With all the previous steps, the Bayesian network is ready to make the prediction based on previous keyword auction history. We evaluate the effectiveness of the prediction using empirical data and the results are presented in Section 4.

3.3 Prediction Model with Website Attraction (Scenario 2)

The previous prediction treat all websites as the same, but in reality websites differentiate from each other by their abilities of attracting traffic, mostly depends on their contents and popularities. Typically the more attractive, the higher volume of visit. The model in this section will enhance the previous model by taking the website attraction factor in to account. This enhance model uses an additional parameter to decide the volume of *CT* and to achieve more accurate prediction.

Obviously visit to any website is time sensitive (i.e., periodic), which means new model needs to consider time slice. Thus, new model considering website attraction is developed and shown as Fig.2. While the main branch in Fig.2 is the same as that of Fig.1, the right one grows up to illustrate the influence of website attraction, which in

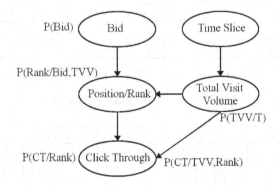

Fig. 2. Bayesian Model for Scenario 2

turn is decided by time slice. This replenishment also results in the change of conditional probability of *CT* from *P(CT|Rank)* substituting for *P(CT|Rank, TVV)* and of *P(Rank|bid)* for *P(Rank|bid, TVV)*respectively. Correspondingly two new probability tables *P(T)* and *P(T|TVV)* are supplemented to calculate the conditional probability above. Surely this model also needs to consider discretization and parameters learning but whatsoever same as that of scenario 1.

We expect that the new model (in Scenario 2) could potentially provide more accurate and reasonable estimated results, whereas it is slightly more complex than the previous one (in Scenario 1). The complication arising from new casual structure, which makes the difficulty of learning and estimating drastically ascend. It is thorny to calculate joint conditional probability in the sense that more states or complicated density function needs to be carefully addressed. Furthermore data for learning will accrue exponentially which may result in a bias estimation of parameters. Fortunately those challenges are fetched in this paper.

4   Empirical Evaluation

4.1 Data Analysis

In order to verify of the prediction from the Bayesian network model developed above we obtained real keyword bidding data from one of the flower vendors in Shanghai in China who has been advertising on Google and Baidu (the largest search engine in China) for several years. An average bid for a day is extracted to constitute the dataset, and the data spans over one year. Statistics of the dataset are summarized in the Tab.1.

Table 1. Statistics of bidding history

Statistics / Variables	Mean	s.d.	Median	Min.	Max.
Bid Price	2. 3360	2. 1366	1. 64	0. 31	13. 35
Rank	12	7. 4419	12	1	24
Click Through	222	157. 2964	166	44	597
Conversion Rate	0. 04899	0. 03949	0. 0415	0	0. 149
Total Clicks	730	168. 9458	713	500	1212

Table 2. Covariance matrix of parameters

Parameters	bid_j	CT_j	$position_j$	TVV_j
bid_i	4.550770	317.7849	-13.45610	94.42269
CT_i	317.7849	24663.86	-1104.836	6744.221
$position_i$	-13.45610	-1104.836	55.20686	-218.6924
TVV_i	94.42269	6744.221	-218.6924	28452.35

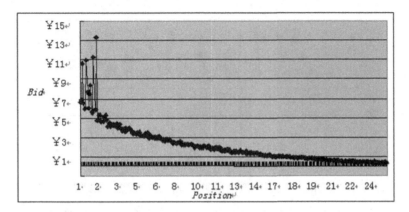

Fig. 3. Illustrates the relationship between position and bid. Just as expected that position inversely correlates to bid.

Tab.2 summarizes the covariance of *bid, CT, position* and *TVV* within two different rounds. One can easily find the negative relationship between *position* and other parameters, including *TVV*. And the smaller value between bid_i and bid_j indicates that new bid heavily depends on history bid while *TVV* seems to be against this conclusion.

4.2 Evaluation Result

Firstly 260 time slices are put along the timeline of bidding for carving up the process. In each slice a couple of data including *bid, position* and *CT* are input as the prior probabilities and a series of posterior probabilities of the three parameters are produced and some rules used by bayesian network model are sequentially worked out. Then other 66 slices as the detector are employed to verify the validity of the rules derived from the training.

Fig.4 shows the true values, predicted values and their gap of *position* given *bid* and *CT* from top to down in scenario 1. Within 66 time slices (horizontal axis) one can find the range (vertical axis) of *position* less than 30. And the last schedule indicates that prediction is rather accurate and the gap between true and predicted values may be nearly ignored.

Table 3. Statistics of prediction for scenario 1

	position	*CT*
Mean	0.878788	−1.51515
Variance	11.03124	2772.362

Tab.3 proves statistics of the gap in scenario 1. It is shown that the mean of *position* gap is very close to zero, which implies that it is determined only by *bid* and website attraction has no effect on this occasion.

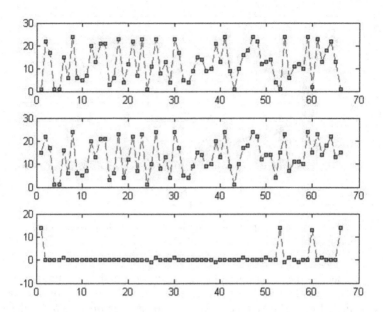

Fig. 4. Original *Position*, Predicted Values and the Differences in Scenario 1

Fig.5 displays the same parameters result for scenario 2 where website attraction and period of time take effect. It is worth of notice that only one more factor is added, though, two joint conditional probabilities need to be deducted and calculated using another dataset. This may result in fluctuation of the prediction.

Similarly Tab. 4 gives the same statistics. Contrary to the expectation, however, mean of the gap in scenario 2 is a little larger than that in scenario 1 and so does variance. The reason can be attributed to the daily data rather than hour even minute data that can indeed reflect the character of clicks on different time point in a day.

Table 4. Statistics of prediction for scenario 2

	position	*CT*
Mean	1. 575758	−1. 51515
Variance	20. 4634	2602. 282

4.3 Application of the Model

Several popular bidding strategies, such as relative position strategy, "gap move" strategy, and cost constraint strategy, are widely applied by those companies, such as KeywordMax, SureHits and Did-It. These companies nested various strategies into their software of advertising campaign management. These strategies, though, do own special advantages respectively the serious weak underlaid them are the local character. Basically those software can manage only one keyword using some strategy while most of advertisers often expect a portfolio management. Besides, the

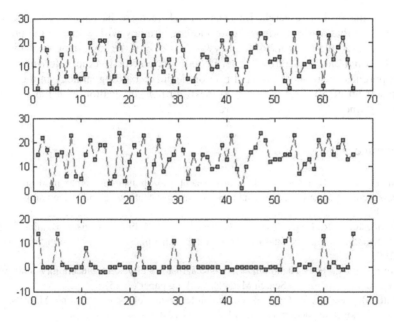

Fig. 5. Original *Position*, Prediction Value and Difference in Scenario 2

strategies above can work either on the special occasions or with passive attribution. So the limitation of those bidding strategies and unsatisfied need stimulate the creation of new strategy.

Obviously Bayesian network as one of the learning models occupies the congenital advantage for its global character. The developed model in this paper definitely does not serve only for one keyword but involve the whole keyword portfolio by providing an adaptive package to manage each keyword for different occasion. Furthermore it may predict bidding result ex ante as well as absorbing history experiences. Therefore one can design the bidding strategy globally using our model and expect to achieve the optimal objectives such as maximal profit. Concretely the main role played by this model is to help estimate parameters of optimization function transformed from the bidding strategy. Thus an automatic adjustment of parameter can ultimately lead to managing each campaign effectively and efficiently.

5 Conclusion

Keywords auction is one of the most significant phenomenon in the current Internet economy. This new business model incurs billions of revenue for search engines but seems to be adverse to lots of advertisers who basically are less knowledgeable about the bidding strategy and competitors behavior. This inappropriate status inspires the research of this paper where two prediction models based on bayesian network are developed. The goal of these models is to help predict bidding result on the occasion of whether or not considering website attraction. Empirical study manifests that the prediction result is close to an unbiased estimation to the true value.

As for the further research, Bayesian model needs to be enhanced to encompass more factors that as the components of bidding strategy will influence the prediction result. For example category of the keywords is one of the important attributions that may affect bid and click through rate. In addition, conversion rate referring a ratio of actual purchase also plays essential role on deciding bidding strategy in the terms of advertisement effect. So we expect to make further verification with more empirical data and refine the estimation structure.

Reference

1. Mandelli, A.: Banners e-mail, advertisement and sponsored search: proposing a value perspective for online advertising. Int. J. Internet Marketing and Advertising 2, 92–108 (2005)
2. Animesh, A., Ramachandran, V., Viswanathan, S.: Online Advertisers Bidding Strategies for Search, Experience, and Credence Goods: An Empirical Investigation. working paper (2005)
3. Animesh, A., Ramachandran, V., Viswanathan, S.: Quality Uncertainty and Adverse Selectio. In: Sponsored Search Markets, working paper (2005)
4. Kitts, B., Leblanc, B.: Optimal Bidding on Keyword Auctions. Electronic Markets, special issue: innovative auction markets 14, 186–201 (2004)
5. Edelman, B., Ostrovsky, M.: Strategic Bidder Behavior in Sponsored Search Auctions. Andreina Mandelli. Banners, e-mail, advertisement and sponsored search: proposing a value perspective for online advertising. Int. J. Internet Marketing and Advertising 2, 92–108 (2005)
6. Parkes, D.C., Sandholm, T.: Optimize and Dispatch Architecture for Expressive Ad Auctions. 05' Electronic commerce. British Columbia, Canada (2005)
7. Ouyang, F.-R., Wang, X.-J., Yang, H.: Agent Bidding Strategy And Simulatio. In: Double Auctions. Proceedings of the 3th International Conference on Machine Learning and Cybernetics, Shanghai, pp. 26–29 (2004)
8. Jensen, F.V.: Bayesian Networks and Decision Graphs. Springer, Heidelberg (2001)
9. Feng, J., Bhargava, H.K., Pennock, D.M.: Implementing Sponsored Search in Web Search Engines: Computational Evaluation of Alternative Mechanisms, working paper (2005)
10. Balcan, M.-F., Blum, A., Hartline, J.D., Mansour, Y.: Sponsored Search Auction Design via Machine Learning. Technical report CMU-CS-05-143 (2005)
11. Nadkarni, S., Shenoy, P.P.: A causal mapping approach to constructing Bayesian networks. Decision Support Systems 38, 259– 281 (2004)
12. Park, S., Durfee, E.H., Birmingham, W.P.: Use of Markov Chains to Design an Agent Bidding Strategy for Continuous Double Auctions. Journal of Artificial Intelligence Research 22, 175–214 (2004)
13. Park, Y.-H., Bradlow, E.T.: An Integrated Model for Bidding Behavior in Internet Auctions: Whether, Who, When, and How Much. Journal of Marketing Research XLII, 470–482 (2005)

Analyzing the Influence of Websites Attributes on the Choice of Newspapers on the Internet

Carlos Flavián and Raquel Gurrea[*]

Department of Economy and Business Studies,
University of Zaragoza,
C/ Gran Vía, nº 2. 50005 (Zaragoza) Spain
Tel.: +34976762719; Fax: +34 976761767
cflavian@unizar.es, Gurrea@unizar.es

Abstract. In the last years, newspaper publishing is one of the activities that have been more affected by the development of the Internet. The appearance of the new digital media has brought far-reaching changes in this sector. This research is one of the first studies that analyse online press reader behaviour on the Internet. Thus, this research studies the key factors that could influence on the choice of digital newspapers. More specifically, this study focuses on the analysis of the effect of usability of the digital newspapers, reputation, trust, privacy and familiarity with the websites, on the final choice of digital dailies. For that, a survey on the Internet was applied. The scales were validated and refined, after which the hypotheses were tested by way of a structural equation model. It was possible to find out that there is an intense effect of usability and familiarity with the websites on the choice of electronic newspaper. However, reputation and trust to the websites do not influence significantly on the choice of digital dailies. In this way, we must note that readers do not perceive risk or costs derived from making wrong choices, because the change of news´ supplier is fast and easy. Also, privacy has not a significant influence on the final reading of digital newspaper. This could be explained by the fact that readers should not usually give personal data to the daily website, except in the case of subscription to special services. In this case, the information is not too private. Consequently, the main aspects which justify digital newspaper reading should be considered by the management in order to potentate its use. Also, newspaper firms should make efforts to improve the levels of usability of their websites. Moreover, digital dailies should develop strategies in order to ensure loyal readers that could be familiarized with the new medium.

Keywords: Digital newspapers, websites, reader behaviour, research paper.

1 Introduction

In last years, the development of the Internet is affecting the way in which most businesses work. Newspaper publishing is one of the activities that have been more

---

[*] The authors are grateful for the financial support received from the Spanish Ministry of Science and Technology (SEC2005-4972; PM34) and the Aragón Goverment (S-46).

G. Psaila and R. Wagner (Eds.): EC-Web 2007, LNCS 4655, pp. 179–190, 2007.
© Springer-Verlag Berlin Heidelberg 2007

affected by the development of the Internet. In fact, the appearance of the new digital media has brought far-reaching changes in this sector [1]. It is possible to note the existence of several changes related to the way in which the editorial staff work and how the news are showed to the readers [2]. The advantages of the new digital media in terms of both supply and demand have brought a revolution in modern journalism. The enormous advantages of the new digital media (speed, distribution costs, immediacy, updating ...) have boosted the service in both quantitative and qualitative terms, resulting in a proliferation of increasingly specialized electronic journals. Indeed, there are currently over 4,200 digital newspapers worldwide [3]. Moreover, digital news and newspapers are among the services most avidly sought by Internet users [4]. News sites and, of course, digital newspapers are among the most widely demanded and visited websites among the Internet community worldwide [5].

In this context, the main aim of this paper is to analyze the key factors that could affect the choice of digital newspapers. More specifically, this research focuses on the study of the effect of usability of the digital newspapers, reputation, trust, privacy and familiarity with the websites, on the final choice of digital dailies.

2 Literature Review Hypotheses

2.1 Reading Digital Newspapers

TAM (*Technology Acceptance Model*) is a relevant model that has been used in many researches in order to analyse consumer behaviour and the adoption of technologies (e.g. 6, 7, 8). In fact, one of the main advantages of this model is its capacity to analyse in depth behavioural aspects, for a wide range of technologies and populations [9]. However, the major critics to the technology acceptance model are focused on the need to include new variables with the aim to improve the knowledge [10]. Also, it seems necessary to adapt TAM to different contexts with special interest [11]. The construct "choice" or "intention to use" is one of the key factors of TAM models. This research explains the final choice of the Internet, as well as the factors affecting this choice, because this topic has not been developed by specialists in press sector.

2.2 Factors Affecting the Reading of Digital Newspapers

Websites usability. Traditionally, the specialized marketing literature has analyzed the importance and influence of consumers' perceptions of establishments and their image regarding actual purchasing attitudes and behavior (e.g. 12, 13, 14).

With the rise of the Internet, a new and innovative store format has appeared. Because of the great interest in this topic and the relative scarcity of literature referring to it, there are many studies that have addressed a new line of research based on the importance of consumer perceptions regarding virtual establishments (e.g. 15, 16) and on the comparison of the most relevant aspects of bricks-and-mortar and digital stores (e.g. 17, 18). In this respect, it seems reasonable to suppose that in the new electronic environment, the design of a website or virtual store might be considered in a very similar way to the concept of image of a traditional establishment. Thus, the design of a website might have a bearing on the likelihood of adopting electronic business. Therefore, it seems reasonable to examine aspects related to website design in more detail [19]. With regard to these aspects, mention should be made of the importance of website usability as a key factor in design and marketing strategy in the new digital economy. Davis [20] defines usability as the degree of effort which, according to the user, is required by the system used – in other words, perceived ease of navigation or purchasing via the Internet. In this way, usability is one of the key factors considered in technologies adoptions [21,22]. So, it seems reasonable to think that if readers perceive adequate levels of usability of newspaper websites, they will be more motivated to the final choice of those websites.

Hypothesis 1: The perceived usability on websites influences positively on the choice of digital newspapers.

Websites reputation. The concept "reputation" refers to the level of equity and honesty of the firm's behaviors [23]. In fact, reputation not only demands doing something because is fair and adequate, but also because it is what must be done [24]. In the digital environment, there is an important scarce of literature related to the influence of websites reputation on consumers' behaviour. Nevertheless, authors like Kollock [25] point out that reputation is a key factor for the success of the transactions on the Internet. In journalistic activities, the previous experience of a user with the firm in the physical channel could affect to the perception of website reputation in the electronic medium. In this way, if readers perceive a high reputation on a newspaper, risk associated to its readership will be reduced. Consequently, the decision making process becomes an easier task.

Hypothesis 2: The perceived reputation on websites influences positively on the choice of digital newspapers.

Trust to websites. The concept of trust has received special attention in marketing literature due to the notable influence it has on the attainment of long lasting and profitable relationships [26]. Traditionally, trust has been defined as a group of beliefs held by the consumer that are derived from perceptions the consumer has about determined attributes that characterise the brand, products or services, salespeople or the establishment [27]. This group of beliefs has been divided into different dimensions and trust is usually considered as a multidimensional construct which differentiates between honesty and benevolence perceived in the behaviour of the second party. At any event, some researchers have contemplated variations in the nature of the dimensions that characterise trust. It is relatively common, for example, to find references to the perceived competence of the second party [28] This latter dimension is concerned with the perceived dexterity and ability of the second party [29,30].

Some researches have analysed the importance of trust in Internet relationships [31,32]. The reason for such interest is the insecurity shown by the consumer when shopping online [33], one of the most important obstacles to the growth of e-commerce [34,35]. In the journalistic business, the firms should look after generating an adequate level of trust among digital dailies' readers, in order to establish relationships in the long time.

Hypothesis 3: The trust to websites influences positively on the choice of digital newspapers.

Privacy on websites. Taking into account the possibilities offered by the new technologies for the processing of communication and information, data privacy is reaching a remarkable importance in the last years. In fact, there is a notable mistrust to the way in which private data are collected and used on the Internet [36,37]. Thus, privacy is one of the main factors to be considered to develop business on the net. In the electronic medium, privacy affects to the obtaining, distribution and unauthorized use of private information [35]. It is necessary to make a correct manipulation of the users' information, in order to overcome the obstacles based on the mistrust [38,39] and therefore, to achieve the final choice and use of the Internet. In fact, recent studies note the existence of a clear influence of the data privacy on the levels of trust showed by individuals [40] and consequently, on consumers' attitudes.

In the case of the newspaper sector, once users show a positive attitude to the reading of digital dailies, the guarantee of privacy of the private data would play an important role on the choice of electronic press.

Hypothesis 4: The perceived privacy on websites influences positively on the choice of digital newspapers.

Familiarity with websites. In the specialized literature, several papers have proved that the users' familiarity with websites would affect to the individuals' final decisions (e.g. 41, 42, 43). Regarding this aspect, authors such as Gefen [44] argue that familiarity reduces the uncertainty and simply the established relationships, as it generates knowledge structures on individuals. In this way, the making decision processes become easier and the necessary cognitive effort is reduced. In this line, Walczuch, Seelen and Lundgren [45] include familiarity as a psychological aspect that affect to the trust to a website. Meanwhile, Bhattacherjee [46] demonstrates the existence of this relationship undoubtedly.

In the journalistic sector, it seems logical to point out that readers familiarized with newspapers websites would have a greater predisposition to read them. In this way, the higher familiarity with the websites, the higher choice of electronic newspapers.

Hypothesis 5: The familiarity with websites influences positively on the choice of digital newspapers.

3 Methodology

3.1 Data Collection

With the aim to develop scales to measure the variables that influence on the reading of digital newspapers, we undertook an exhaustive literature review and searched for existing scales that could be suitably adapted. Having established the context of the analysis and designed a series of questions that would pick up the various factors involved, the information contained in the preliminary scales was presented to a group of academic and business experts for their remarks on both formal and conceptual issues. The survey was conducted through the Internet. The questionnaire was published on a website designed specifically for the purposes of our research. Information was sent to various distribution lists and banners were placed on the websites of electronic newspapers in order to disseminate the existence of our research.

The total number of valid responses obtained in the period April-July 2006 was 239 (readers of electronic newspapers written in Spanish language). The representative nature of the sample could be guaranteed by the similarity of the profile of the interviewees with that usually obtained in recent studies about the Internet users, such as AECE [37] or AIMC [47]. More specifically, the majority of the interviewees were male, aged between 25 and 34, a high level of education and over 5 years' experience with computers and Internet use.

3.2 Reliability and Validity Analysis

The first step in assessing reliability was to calculate the Cronbach's alpha value [48] and the item-total correlation of each of the variables analyzed [49]. The results of Cronbach's alpha test showed an acceptable degree of internal consistency in the eight scales considered, being in all the cases over the 0.7 recommended by Cronbach [48] or Nunnally [50]. The Reading digital newspapers (READ) scale reached .90 points; Usability (USAB), .91; Reputation (REP), .98; Honesty (HON), .85; Benevolence (BEN), .89; Competence (COM), .88; Privacy (PRIV), .95 and Familiarity (FAM), .76. In addition, the item-total correlation of each indicator was higher than the 0.3 points recommended by Nurosis [51]. As a second stage in the previous exploratory analyses of the scales, we carried out a study of unidimensionality in each by means of an exploratory factor analysis of the principal components, and where necessary, with a varimax rotation [52]. The number of factors extracted through the eigenvalue criteria was 1; there was always a high variance results: READ 78.77%; USAB 69.23%; REP 97.20%; HON 69.82%; BEN 70.21%; COM 81.48%; PRIV 85.18%; and FAM 69.09%. Furthermore, the factor loadings were above 0.5 points [53].

In order to guarantee the scales' reliability and validity, we carried out a series of confirmatory analyses, according to the methodology of confirmatory model development [53]. This methodology enables one to sift scales by the development of successive confirmatory factor analyses[1]. With this aim, we successively eliminated those items which did not meet the three criteria proposed by Jöreskog and Sörbom [54][2]. Thus, it was necessary to eliminate the following items: READ4, USAB6, FAM3 and FAM4. Once the scales were depurated, the required criteria were met and the structural model fit was acceptable. Also, we carried out tests of the composite reliability coefficient[3] [55] and the average variance extracted[4] [56]. With the aim of contrasting the presence of a multi-dimensional structure in the multi-dimensional

---

[1] We used the statistical software EQS version 6.1.

[2] 1) Criteria of weak convergence would eliminate indicators that did not have a significant factorial regression coefficient (t student>2.58: p=0.01).

    2) Criteria of strong convergence would eliminate those indicators that were not substantial, i.e. those whose standardized coefficient is less than 0.5.

    3) Lastly, Jöreskog and Sörbom[54] propose the elimination of those indicators that least contribute to the explanation of the model, considering the cut-off point as $R^2<0.3$.

[3] READ, 0.93; USAB, 0.92; REP, 0.99; HON, 0.86; BEN, 0.92; COM, 0.86; PRIV, 0.96; and FAM, 0.76.

[4] READ, 0.83; USAB, 0.68; REP, 0.96; HON, 0.61; BEN, 0.69; COM, 0.62; PRIV, 0.81; and FAM, 0.63.

construct (trust), a rival model strategy was developed [57]. The results showed a higher fit in the second order model which allowed us to confirm the multidimensionality of the trust variable.

No statistical tests exist that would allow us to guarantee the validity of the scale contents. In general, the validity of a scale's content may be accepted if it has been developed on the basis of existing theories in the relevant literature. In our case, we may consider validity to be guaranteed not only in view of the rigor employed in the design of the initial scales on the basis of the literature (e.g. 58, 59, 60, 61, 33, 23, 62, 44), but also because we have taken into account the comments of various experts in the subject. Besides, once the scales were revised according to the experts' comments, it was carried out a pretest.

Construct validity analysis is formed by two fundamental categories of validity:

- Convergent validity: The standardized coefficients in each scale were over 0.5 and significant [63]
- Discriminatory validity: To assess the discriminatory validity we used several tests, such as checking that value '1' was not in the confidence interval of the correlations between different constructs. Likewise, we checked that correlations between different scales were not over 0.8 [64].
- Finally, the correlation between each pair of variables was set at "1" and we conducted a χ^2 differences test, in order to test that the initial model was significantly different (p<0.001) to any of the alternative models [65].

On the basis of preliminary analysis, we were able to establish the definitive scales to measure each of the variables considered in this study (see Appendix 1).

4 Structural Model Analysis

Once the measurement scales were designed and validated, we proceeded to contrast the different hypotheses which made up the structural model of analysis (see figure 1). In the case of the multi-dimensional variable (trust), the indicators that were taken for causal analysis were derived from the arithmetical average of the items that composed each of the dimensions [66,67].

It was noted that usability influences on the final choice of digital newspapers. Therefore, it was not possible to reject hypothesis H1. Also, we proposed the possible existence of positive effects of reputation and trust on reading electronic dailies. However, it was not possible to demonstrate these relationships. On the contrary, the intensity of these no significant effects is very light. So we should reject hypotheses H2 and H3. These obtained results could be considered surprising. It seems reasonable that aspects such as reputation or trust to the newspapers play a remarkable role in the reading decision. However, a good number of readers declare that the consult of current news does not involve risk, because it is possible to correct a mistake easily, only visiting another newspaper website. This ease of comparison involves a smaller perceived risk and consequently, the role played by factors related to guarantee a correct choice (e.g. reputation, trust) is not significant.

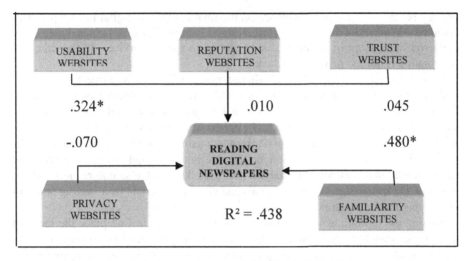

Fig. 1. Structural equation model. Standarized solution

On other hand, it was not possible to accept the hypothesis H4. We did not find evidences for the relationship between the privacy and the choice of digital press for reading news. A plausible explanation for this result would be based on the fact that, although the privacy of data is a key aspect in electronic commerce on the net, the journalistic business presents some peculiarities that could low its relevance. In fact, users fear to give private or banking data, in general. But, excepting the case of subscription to special services, newspapers hardly request information. Anyway, the requested data are scarce, exactly the opposite that occurs in commercial websites. Finally, it was possible to corroborate the existence of a clear relationship between familiarity with the websites and the choice of them. Consequently, we must accept hypothesis H5. The high R^2 obtained ($R^2 = 0.438$) should be noted, as well as the good overall fit of the structural model (e.g. RMSEA=0.069; NFI=0.865; NNFI=0.852; CFI=0.885; Normal χ^2=2.1)

5 Conclusions, Managerial Implications, Limitations and Future Research

Newspaper sector is one of the most influenced activities by the development of the digital channel. This research takes as a point of reference the need of analysing in depth digital newspapers readers' behaviour. More specifically, there is an important lack of literature and it is interesting to study the key factors that affect the final decision of reading news on the Internet. The empirical analysis has allowed us to note a clear and intense effect of usability and familiarity with the websites on the choice of electronic newspaper for reading current news. However, reputation and trust to the websites do not influence significantly on the choice of digital dailies. In this way, we must note that readers do not perceive risk or costs derived from making wrong choices, because the change of news´ supplier is really fast and easy. Also, privacy has not a significant influence on the final reading of digital newspaper. This could be explained by the fact that readers should not usually give personal data to the daily website, except in the case of subscription to special services. In this case, the information is not too private or delicate.

Thus, newspaper editorial firms should make efforts to improve the levels of usability of their websites. Thus, users could perceive the reading of news on the Internet as an easy and comfortable task that let them have a good time. In the same way, digital dailies should develop and implement strategies in order to ensure loyal readers that could be familiarized with the new medium. This fact would motivate the consultation of current information in the electronic channel. Besides, journalistic firms should make efforts to analyze and gain knowledge about readers' needs. Anyway, the knowledge of readers' goals and preferences could be a starting point for developing marketing strategies.

Among the limitations of our study, we would mention in first place the fact that the sample is made up from Spanish language newspapers´ readers. Nevertheless, the relevance of the sample could be guaranteed because the Spanish language is the fourth language used on the Internet [5]. Nevertheless, it could be convenient to validate the model with a wider and international sample. Also, it would be interesting to analyze the effects of usability and trust to digital newspapers, on readers' satisfaction [61]. Finally, it would be interesting to analyze some moderating effects in the model. Specifically, it could be convenient to include the moderating effect of Internet user experience. Thus, it seems reasonable to think that readers with a greater level of Internet experience show a deeper knowledge of the digital channel.

References

1. Bush, V., Gilbert, F.: The web as a medium: an exploratory comparison of Internet users Versus Newspapers readers. Journal of Marketing Theory and Practice 10(1), 1–10 (2002)
2. Greer, J., Mensing, D.: The evolution of Online Newspapers: A longitudinal content analysis, 1997-2003. In: Newspaper division of the Association for Education in Journalism and Mass Communication for the 2003 annual conference (2003)
3. Editor & Publisher. Newspapers web sites continue to gain readers (2005) www.editorandpublisher.com/eandp/departments/online
4. Levins, H.: Growing U.S. audience reads news on net. Editor & Publisher 131(8), 14 (1998)
5. Newspaper Association of America (2003) www.naa.org
6. Lederer, A., Maupin, D., Sena, M., Zhuang, Y.: The technology acceptance model and the World Wide Web. Decision Support Systems 29(3), 269–282 (2000)
7. Gefen, D., Straub, D.W.: Consumer trust in B2C e-commerce and the importance of social presence: experiments in e-products and e-services. OMEGA: The International Journal of Management Science 32, 407–424 (2004)
8. Shih, H.: Extended technology acceptance model of Internet utilization behaviour. Information & Management 41, 719–729 (2004)
9. Lee, M., Cheung, C., Chen, Z.: Acceptance of Internet-based learning medium: the role of extrinsic and intrinsic motivation. Information & Management 42, 1095–1104 (2005)
10. Legris, P., Ingham, J., Collerette, P.: Why do people use information technology? A critical review of the technology acceptance model2. Information & Management 40, 191–204 (2003)

11. Hartwick, J., Barki, H.: Explainig the role of user participation in information system use. Management Science 40(4), 440–465 (1994)
12. Lindquist, J.: Meaning of image: a survey of empirical and hypothetical evidence. Journal of Retailing 50(4), 29–37 (1974)
13. Ghosh, A., Chakraborty, G., Ghosh, D.: Improving brand performance by altering consumers brand uncertainty. Journal of Product and Brand Management 5(5), 14–20 (1995)
14. Erdem, O., Oumlil, A., Tunclap, S.: Consumer values and importance of store attributes. International Journal of Retail and Distribution Management 27(4), 137–144 (1999)
15. Raijas, A.: The consumer benefits and problems in the electronic grocery store. Journal of Retailing and Consumer Services 9(2), 107–113 (2002)
16. Sim, L., Koi, S.: Singapore's internet shoppers and their impact on traditional shopping patterns. Journal of Retailing and Consumer Services 9(2), 115–124 (2002)
17. Loshe, G., Spiller, P.: Internet retail store design: how the user interface influences traffic and sales. Journal for Computed-Mediated Communication 5(2) (1999) http://www.ascusc.org/jcmc/vol5/issue2/lohse.htm
18. Dahlén, M., Lange, F.: Real consumers in the virtual store. Scandinavian Journal of Management 18, 341–363 (2002)
19. Fogg, B., Kameda, T., Boyd, J., Marshall, J., Sethi, R., Sockol, M., Trowbridge, T.: Stanford-Makovsky Web Credibility Study 2002: Investigating what makes Websites credible today. A Research Report by the Stanford Persuasive Technology Lab & Makovsky & Company. Stanford University (2002) http://www.webcredibility.org
20. Davis, F.: Perceived usefulness, perceive ease of use and user acceptance of information technology. MIS Quarterly 13(3), 319–329 (1989)
21. Straub, D.W., Limayem, M., Karahanna, E.: Measuring system usage: implications for IS theory testing. Management Science 41(8), 1328–1342 (1995)
22. Hu, X., Lin, Z., Zhang, H.: Myth or reality: Effect of trust promoting seals in electronic markets. In: WITS, New Orleans, LA, USA (1999)
23. Doney, P., Cannon, J.: An Examination of the Nature of Trust in the Buyer-Seller Relationship. Journal of Marketing 51, 35–51 (1997)
24. Guinalíu, M.: La gestión de la confianza en Internet. Un factor clave para el desarrollo de la Economía Digital. Unpublished doctoral tesis, University of Zaragoza (Spain) (2005)
25. Kollock, P.: The production of trust in online markets (1999) http://www.sscnet.ucla.edu
26. Hunt, S., Morgan, R.: The Commitment-Trust Theory of Relationship Marketing. Journal of Marketing 58, 20–38 (1994)
27. Ganesan, S.: Determinants of Long-term Orientation in Buyer-Seller Relationship. Journal of Marketing 58, 1–19 (1994)
28. Sako, M., Helper, S.: 2Determinants of trust in supplier relations: Evidence from the automotive industry in Japan and the United States. Journal of Economic Behaviour and Organization 34(3), 387–417 (1997)
29. Mayer, R., Davis, J., Shoorman, F.: An Integrative Model of Organizational Trust. Academy of Management Review 20(3), 709–734 (1995)
30. Sirdeshmukh, D., Singh, J., Sabor, B.: Consumer trust, value and royalty in relational exchanges. Journal of Marketing 66, 15–37 (2002)
31. McKnight, D., Chervany, N.: What trust means in e-commerce customer relationships: an interdisciplinary conceptual typology. International Journal of Electronic Commerce 6(2), 35–59 (2002)
32. Lee, J., Kim, J., Moon, J.: What makes Internet users visit cyber stores again? Key design factors for customer loyalty. In: Proceedings of the Conference on Human Factors in Computing Systems CHI 2000, pp. 305–312. ACM Press, New York (2000)
33. Jarvenpaa, S., Tractinsky, M., Vitale, M.: Consumer Trust in an Internet Store. Information Technology and Management 1(1-2), 45–71 (2000)
34. Korgaonkar, P., Wolin, L.: A multivariate analysis of web usage. Journal of Advertising Research 39, 53–68 (1999)

35. Wang, H., Lee, M., Wang, C.: Consumer privacy concerns about Internet marketing. Communications of the ACM 41, 63–70 (1998)
36. Udo, G.: Privacy and Security concerns as major barriers for e-commerce: A survey study. Information Management & Computer Security 9(4), 165–174 (2001)
37. Asociación Española de Comercio Electrónico, AECE (2005) www.aece.org
38. Lardner, J.: I Know what you did last summer and fall. US News & World Report 126(15), 55 (1999)
39. Hoffman, D.L., Novak, T.P.: Marketing in Hypermedia Computer-Mediated Environments: Conceptual Foundations. Journal of Marketing 60(3), 50–68 (1996)
40. European Comisión.: Segues relating to Business and Consumer eCommerce. Special Eurobarometer 60.0/Wave 201. European Opinion Group (2004)
41. Bettman, J.R., Park, C.W.: Effects of prior knowledge and experience and phase of the choice process on consumer decision processes. A protocol analysis. Journal of Consumer Research 7, 234–248 (1980)
42. Ratneshwar, S.A., Shocker, D., Stewart, D.W.: Toward understanding the attraction effect. The implications of product stimulus meaningfulness and familiarity. Journal of Consumer Research 13, 520–533 (1987)
43. Gefen, D., Straub, D.W: Gender differences in the perception and use of email: an extension to the technology acceptance model. MIS Quarterly 21(4), 389–400 (1997)
44. Gefen, D.: E-commerce: the role of familiarity and trust. OMEGA: The International Journal of Management Science 28, 725–737 (2000)
45. Walczuch, R., Seelen, J., Lundgren, H.: Psychological determinants for consumer trust in e-retailing. In: Proceedings of the VIII Research Symposium on Emerging Electronic Markets (2001) http://www-i5.informatik.rwth-achen.de/conf/rseem2001/papers/walczuch.pdf
46. Bhattacherjee, A.: Individual trust in online firm: scale development and initial test. Journal of Management Information Systems 19(1), 211–241 (2002)
47. Asociación para la Investigación de Medios de Comunicación, AIMC (2006) www.aimc.es
48. Cronbach, L.J.: Essentials of psychological testing. Harper and Row, New York (1970)
49. Bagozzi, R.P.: Evaluating Structural equations Models with Unobservable Variables and Measurement Error: A Comment. Journal of Marketing Research 18, 375–381 (1981)
50. Nunnally, J.C.: Psychometric Theory, 2nd edn. Mcgraw-Hill, New York (1978)
51. Nurosis, M.: SPSS. Statistical Data Analysis, SPSS Inc. (1993)
52. McDonald, R.P.: The Dimensionality of Test and Items. British Journal of Mathematical and Statistical Psychology 34, 100–117 (1981)
53. Hair, J.F., Anderson Jr., R.E., Tatham, R.L., Black, W.C.: Multivariate Analysis. Prentice Hall, Englewood Cliffs (1999)
54. Jöreskog, K.G., Sörbom, D.: LISREL 8: Structural Equation Modeling with the SIMPLIS Command Language. Ed. Scientific Software International, Chicago-Illinois (1993)
55. Jöreskog, K.: Statistical analysis of sets of congeneric tests. Psychometrika 36, 109–133 (1971)
56. Fornell, C., Larcker, D.: Structural Equation Models With Unobserved Variables and Measurement Error. Journal of Marketing Research 18, 39–50 (1981)
57. Steenkamp, J., Van Trijp, H.C.M.: The use of LISREL in validating marketing constructs. International Journal of Research in Marketing 8, 283–299 (1991)
58. Wu, J., Wang, S.: What drives mobile commerce? An empirical evaluation of the revised technology acceptance model. Information & Management 42, 719–729 (2005)
59. Lai, V., Li, H.: Technology acceptance model for internet banking: an invariance analysis. Information & Management 42, 373–386 (2005)
60. Wang, Y., Wang, Y., Lin, H., Tang, T.: Determinants of user acceptance of Internet banking: an empirical study. International Journal of Service Industry Management 14(5), 501–519 (2003)
61. Flavián, C., Guinalíu, M., Gurrea, R.: The Role played by perceived usability, satisfaction and consumer trust on website loyalty. Information and Management 43, 1–14 (2005)

62. Janda, S., Trocchia, P., Gwinner, K.: Consumer perceptions of Internet retail service quality. International Journal of Service Industry Management 13, 412–431 (2002)
63. Sanzo, M., Santos, M., Vázquez, R., Álvarez, L.: The Effect of Market Orientation on Buyer-Seller Relationship Satisfaction. Industrial Marketing Management 32(4), 327–345 (2003)
64. Bagozzi, R.P.: Structural Equation Model in Marketing Research. Basic Principles. In: Principles of Marketing Research, pp. 317–385. Blackwell Publishers, Oxford (1994)
65. Bagozzi, R., Yi, Y.: On the Evaluation of Structural Equation Models. Journal of the Academy of Marketing Science 16(1), 74–94 (1988)
66. Bentler, P.M.: EQS Structural Equations Program Manual. Multivariate Software, Inc., Encino, CA (1995)
67. Jaccard, J., Wan, C.K.: Lisrel approaches to interaction effects in multiple regression. Sage Publications, Thousand Oaks, CA (1996)

Appendix 1: Definitive Scales of Measurement

The individual is asked to grade from 1-7 their level of agreement or disagreement with the following statements:

SCALE READING DIGITAL NEWSPAPERS	
READ1	I usually read digital newspapers
READ2	I frequently read news on the Internet
READ3	I read press on the Internet quite often
READ4	*I have decided to read news on the digital newspapers*
SCALE USABILITY	
USAB1	It is easy to use this digital newspaper, even when using it for the first time
USAB2	It is easy to find the information I need from this digital newspaper
USAB3	The structure and contents of this digital newspaper are easy to understand
USAB4	It is easy to move within this digital newspaper
USAB5	In this digital newspaper, everything is easy to understand
USAB6	*I feel that I am in control of what I can do when I am navigating this digital newspaper*
SCALE REPUTATION	
REP1	This digital newspaper has a good reputation
REP2	This digital newspaper has a good reputation compared to other electronic dailies
REP3	This digital newspaper is supposed to offer good information and services
REP4	This digital newspaper is supposed to be fair in relationships with users
SCALE TRUST	
	HONESTY
HON1	I think that this digital newspaper usually fulfils the commitments it assumes
HON2	I think that the information offered by this digital newspaper is sincere and honest
HON3	I think that I can have confidence in the promises that this digital newspaper makes
HON4	This website is characterized by the frankness and clarity of the services that it offers

	BENEVOLENCE
BEN1	I think that this website is concernid with the present and future interests of its users
BEN2	I think that this website takes into account the repercussions that their actions could have on the user
BEN3	I think that this website would not do anything intentional that would prejudice the user
BEN4	I think that the design and offer of this website take into account the desires and needs of its users
BEN5	I think that this website is receptive to the needs of its users
	COMPETENCE
COM1	I think that this website has the necessary abilities to carry out its work
COM2	I think that this website has the sufficient experience with its offer
COM3	I think that this website has the necessary resources to successfully carry out its activities
SCALE FAMILIARITY	
FAM1	I am quite familiarized with this digital newspaper
FAM2	I am quite familiarized with the services this digital newspaper offers.
FAM3	*I am quite familiarized with other digital newspapers´ websites use*
FAM4	*I am regular user of digital press websites´ services*
SCALE PRIVACY	
PRIV1	I think that this website is concerned about its users´ privacy
PRIV2	I feel safe when sending personal information to this website
PRIV3	I think that this website is respectful with data protection laws
PRIV4	I think that this website just gather personal data needed for its activities
PRIV5	I think that this website respects users´ data protection rights
PRIV6	I think that this website will not offer my personal data to other businesses without my permission

Note: Items in italic were eliminated on refining process.

Impact of Web Experience on e-Consumer Responses*

Carlota Lorenzo[1], Efthymios Constantinides[2], Peter Geurts[3], and Miguel A. Gómez[1]

[1] University of Castilla-La Mancha. Faculty of Business. Plaza de la Universidad, 1. 02071, Albacete, Spain
Carlota.Lorenzo@uclm.es; MiguelAngel.GBorja@uclm.es
[2] University of Twente. School of Management and Governance, Department NIKOS
P.O. Box 217. 7500 AE Enschede, The Netherlands
E.Constantinides@utwente.nl
[3] University of Twente. School of Management and Governance, Departement of Research Methods and Statistics. P.O. Box 217. 7500 AE Enschede, The Netherlands
P.A.T.M.Geurts@utwente.nl

Abstract. Identifying the Web Experience components and understanding their role as inputs in the online customer's decision-making process is the first step in developing and delivering an attractive online presence, likely to have the maximum impact on Internet users. Based on background literature, this study is focused on the effects of five web experience factors on virtual buying behaviour, specifically, on the choice of a virtual vendor. Additionally, in the model two user behavioural variables –motivation and experience – have been included in order to analyze their impact on web experience elements and, in consequence, on the choice of online vendors. An online consumer survey was held in a realistic virtual shopping environment. The main results of the empirical study show that four of the five web experience components analyzed (i.e., usability, trust-building, marketing mix, and aesthetics) have a positive and significant effect on the choice of e-vendor while interactivity does not have any significant effect in this choice. Experience is also an influential variable while buying motives do not substantially affect the online customer behaviour.

Keywords: Web experience, virtual buying behaviour, online marketing tools, online marketing.

1 Introduction

The ever-increasing importance and role of the Internet as a commercial tool and marketing channel has put businesses under pressure to professionalize their online activities. The Internet has gained ground against traditional media and has led to increased customer empowerment and sophistication [1].

There is plenty of evidence that the large majority of wired consumers consider by now the web as their primary source of information when searching for products,

* This study was developed within two Research Projects: Research Project whose reference is PCI-05-017, Plan Nacional de Investigación Científica, Desarrollo e Innovación Tecnológica, Junta de Comunidades de Castilla-La Mancha, 2005-2007; and Research Project whose reference is TC20070056, Vicerrectorado de Investigación, Universidad de Castilla-La Mancha, Spain, 2007.

G. Psaila and R. Wagner (Eds.): EC-Web 2007, LNCS 4655, pp. 191–200, 2007.
© Springer-Verlag Berlin Heidelberg 2007

services, news, weather, travel directions or entertainment: according to a recent BurstMedia survey 57.1% of all US web users over 18 years old use the Internet as their primary source of information about products and services they intend to buy[1]. This percentage is even higher (69.2%) among the most affluent consumers, with income of $75.000 or more. In the background of these developments it is not surprising that marketers increase their efforts to attract audiences to their web sites, something evident by the substantial increase of online marketing budgets. A study by TSN Media Intelligence[2] found that in the first quarter of 2006 advertising spending on the Internet –excluding paid search advertising– showed the highest increase (19.40%) against all other media and projected that 12% of the total advertising spending in the US will be spend in 2006 in online advertising, an increase by 13% against 2005. With the Internet becoming the main information source and a major distribution channel the task of attracting customers to the company web site becomes a strategic imperative. Presenting web users with superb online experience is necessary for attracting the attention of the virtual customers and persuading them to engage in online business.

The main objective of this paper is to analyze the impact of web experience on virtual buying behaviour. Based on the background literature as a starting point, we propose a model to study the signification of the constructs analyzed. Moreover, a number of internal variables have been included in the model in order to observe their influence on web experience.

2   Literature Frame: The Virtual Experience

The customer experience from visiting a web site has been recognized as one of the most important factors for online success. According to [2], "creating a compelling online experience for cyber customers is critical for creating competitive advantage on the Internet". The same authors argue that relatively little is known about the factors that contribute to a superb online experience noticing that "online marketers need to develop a comprehensive understanding of consumer behaviour in commercial online environments".

During the last six years a substantial amount of academic research has been carried out with the purpose of understanding the online consumer behaviour. The propensity of consumers to engage in online business has been extensively studied and analyzed [e.g. 3; 4; 5; 6; 7; 8; 9; 10; 11; 12; 13; 14]. Many researchers emphasize that the quality of the online presence is an important influencer of the online consumer's behaviour, something regularly confirmed in research conducted by non-academic parties [15].

More recently researchers have focused their attention on the effects of shopping enjoyment on online consumer satisfaction [16] and on the browsing behaviour as background for effective Website design [17]. Several aspects on the customer experience and particularly on the visual aspects of Web sites have also attracted the attention of researchers [e.g. 18; 19; 20; 21; 22; 23; 24; 25]. [26] suggested that the

---

[1] MarketingVox, April 20, 2006.
[2] TSN Media Intelligence, http://www.tns-mi.com/news/05312006.htm

visual impression of web sites is very important for forming a positive or negative opinion about the quality of a web site and an exposure of 50 milliseconds to a web site is enough for most people to form this opinion. A similar study of [27] has also proven the consistency of the immediate aesthetic impressions.

While an aesthetically appealing web site is the basic requirement for attracting virtual customers, visual attractiveness is one of several elements that combined shape the Online (or Web) Experience. The Web Experience (WE) can be defined as "the total impression online customers get about the virtual firms" [28] as "the result of exposure to a combination of notions, emotions and impulses caused by the design and other marketing elements of the online presentation" [29]. As such the WE is influenced by factors like searching, browsing, finding, selecting and evaluating information as well by impressions generated during interaction and transaction with the online firm.

[2] based on a conceptual model of flow describing the components of "a compelling online experience" [30], concluded that it is possible to define its ingredients, to measure them and relate them to important marketing variables. Other researchers have applied the flow theory [31] as the framework of analysis of human-computer interaction and as a model describing different aspects of the online consumer's behaviour [32; 33]. For all intents and purposes the large number of variables affecting the WE and the constantly changing, dynamic character of the online environment underline the need for more research on the components of the WE and continuous refinement of business approaches [34].

[29], based on a literature review, identified the different elements of the online experience and clustered them in three categories.

a. Content category: Factors exercising a direct and powerful influence on the WE by making the website aesthetically positive and its offer tangible and attractive. They include the Aesthetics and Marketing Mix factors.

b. Psychological category: Web sites must communicate trust and ensure users of the vendor's integrity and credibility in order to persuade customers to stop, explore them, and interact online. Building trust is possible by deploying uncertainty-reducing elements, ensuring the safety of customer personal information and transaction data, eliminating fears of fraud and building trust between the online user and the often unknown and far away located vendor.

c. Functionality category: Factors enhancing the online experience by presenting the virtual client with a good functioning, easy to use and search as well as interactive web site. The Functionality category includes the Usability and Interactivity factors.

This classification was the basis of an empirical study in The Netherlands meant to identify the relative importance of the different factors as influencers of online consumer's decision making process [35].

3 Research Hypotheses

Websites presenting their customers with superb WE not only meet the users' needs, augment their expectations and emotions, but also offer assortment, security, right goods and services, etc. [36]. The combination and the nature of different web experience factors present in a website result in different types of online store designs which cause different perceptions in users which, in turn, affect their shopping behavior [35]. Moreover, the internal variables consumer (i.e. involvement, motivation, experience, and so on) affect their purchase process and final decision [e.g. 37; 29; 38]. On the basis of the above stream we propose the following hypotheses:

H1: The Web Experience Factors are significant influencers of the online buyers' preferences.

H1a: The Usability factor is significant influencer of the online buyers' preferences.

H1b: The Interactivity factor is significant influencer of the online buyers' preferences.

H1c: The Trust factor is significant influencer of the online buyers' preferences.

H1d: The Aesthetics factor is significant influencer of the online buyers' preferences.

H1e: The Marketing Mix factor is significant influencer of the online buyers' preferences.

Some authors suggest that usability reflects the perceived ease and usefulness for the navigation through the Internet [e.g. 37; 22; 39]. Other studies found that usability is a very important attribute for achieving desirable internal and behavioural responses [e.g. 8; 19; 40; 41; 42]. The Marketing Mix elements are widely considered as the main controllable influencers of consumer behaviour [e.g. 43; 44; 45; 46]. Nevertheless, the introduction of the Internet as a business management element and as the main interface with the customer has questioned the importance of the Marketing Mix elements as the main influencers of the online consumer [47]. In this sense it is important to understand the significance of the new elements of influence and their relevant importance versus the traditional Marketing Mix. We propose the following hypothesis:

H2: Online customers prefer to buy from web shops scoring better in Usability and Trust while the Marketing Mix is not the main influencer of the online buying preference.

The personal attributes of consumers (i.e. involvement, motivation, experience, ability to Internet adaptation, and so on) affect their purchase process and final decision [e.g. 37; 19; 38]. Based on [35], and in order to analyze the influence of two specific internal variables (i.e. motivation and experience) on users' preferences, we propose two hypotheses, as follows:

H3: The motives of online customers to buy online do not have an effect on the way the WE factors influence their online vendor preference.

The next hypothesis tests (a) the effect of the user's affinity with the Internet expressed in the number of years one is using it and (b) the effect of the user previous experience with online purchases. So finally, in this study is proposed that:

H4: The degree of experience of virtual customers in online shopping affects the importance they attribute to each Web Experience factors (Usability, Interactivity, Trust, Aesthetics, and Marketing Mix) as influencers of their online vendor decisions.

H4a: The number of years one is using the Internet affects the importance that e-buyers attribute to WEF as influencers of their online vendor decisions.

H4b: The experience with the online purchase affects the importance that e-buyers attribute to WEF as influencers of their online vendor decisions.

4 Methodology

4.1 The Scenario

Participants in the survey were recruited from the student ranks of a research university in Spain. The study was conducted in a realistic virtual shopping environment in the computer laboratory where users were instructed by supervisors on how to carry out an online shopping assignment (i.e. searching and buying online a new –not second-hand– digital camera with a certain technical characteristics, within online stores destined to that objective –not auction websites such as e-Bay–) and fill in a number of questionnaires. The questionnaire was available online and was divided in two sections. The first one (i.e. "Introduction form") included questions about basic demographics and questions about the users' attitudes towards online shopping and previous experience with the Internet. The second section contained three forms: A, B, and C. In form A the participants had to indicate their experiences about the online store where they bought the camera; in form B they had to record their experience from the online shop of their second choice, a web shop they found attractive enough and saw as alternative option. In Form C they had to indicate their experiences for a virtual shop that they found unattractive and they would never choose for an online purchase.

The final sample was composed of 204 participants, divided in nine sessions. The procedure consisted on exposing to people to virtual shopping experience under specific instructions commented at the beginning of the experiment by an instructor. Once users had filled the introduction questionnaire, they had to search and buy a digital camera through the Internet. During their online visit users had to compose two lists of vendors: "Favourite" and "No favourite" web sites according to their experienced sensation within them. A fictitious amount of 300 Euros per participant (including price and postal costs) was available to spend on purchasing a digital camera. Based on [35], the time to carry out the virtual search and purchase was limited to 30 minutes.

4.2 Measurement of Variables

Once finished the purchase, the user had to fill the rest of online survey (A, B, and C forms). The web-based tool developed for this research included an automatic tracking process based on e-agent software to track and record all click-throughs and times related to the browsing behaviour during the experiment in order to obtain information on the type of websites visited, times spent in each website and each section within the website, and so on. Each part of online survey contained different types of variables [35].

The form I included questions about demographic characteristics and questions about the users' attitudes towards online shopping and previous experience with the Internet (buyers/no buyers), asking them the three main motives for shopping or not shopping online.

Within A, B, and C forms were included questions -five points Likert scale- related to users' perceptions of each website (as explained above) in every one of the 25 individual characteristics making up the five factors of the WE (e.g. "It is convenient to buy products in this online shop", "the shop offers excellent customer service", "the site offers adequate guarantees for the safety of online transactions", "the site's design is superb", "the site offers a wide deep product assortment", etc.).

5 Results

5.1 Main Descriptive Data

Most of the users were female (63%); the majority of them between 18 and 22 years old. An important percentage of participants (73%) were experienced Internet users with more than two years of active Internet usage, although only 31% of them had previously bought goods or services online; 15% of those belonging to the last category spend between 50 and 100 Euros per year for online purchases. Moreover, 83% of the participants had a credit card.

For the majority of those who do not buy products in the Internet (25%) the most important reason for that was that they prefer shopping in other ways while 21% mentioned the lack of physical contact with the product as an impeding factor. The ease of finding better prices (21%) and comparing prices (20%) were the first and second more frequently mentioned main reasons for online shopping. Twenty five different brands of digital cameras (e.g. Sanyo, Olympus, Canon, Fujifilm, Benq, Samsung, Sony, HP, and so on) were "bought" by users, including a large variety of models.

Finally in the "Introduction Form", participants were asked to indicate in a five-points Likert scale their impression as to the importance of the five WE factors on their choice of an online vendor in order to observe the users' perceptions versus the actual behaviour. In general, the participants consider all five the WE elements are relevant influencers of their choice for online shops (fourth point of Likert scale), although especially the Trust and Marketing Mix elements (fifth point of Likert scale).

5.2 Statistical Data

The three websites chosen by users have been analyzed from the responses on 25 WE elements where participants could (totally) agree, neither agree nor disagree, or (totally) disagree. In order to test our hypotheses, a factorial analysis was carried out in order to reduce the number of WE items. As result, we obtained five factors. An overview of measurement of web experience factors is showed in table 1.

Table 1. Measurement of WE components

Factor Label	Example question	Number of indicators	First Eigen Value	Cronbach Alphas
Usability	Q3. Information easily accessible	7	3.57	.83
Interactivity	Q8. Excellent customer service	4	2.04	.67
Trust	Q16. Transparent guarantee policy	5	2.85	.81
Aesthetics	Q18. High presentation quality	4	2.59	.81
Mk Mix	Q24. Very competitive prices	5	2.61	.77

A binomial logistic regression was executed with the five WE elements per website as independent variables and the purchasing behavior as the dichotomy explained variable (i.e. buy/not buying). In table 2 are showed the users' buying preferences regarding the WE factors.

Table 2. Impact of WE factors on e-consumer preferences

Dependent variable (buy/not buying)		E-consumer preferences		
Hypothesis		H1/H2	H3	H4a/b
Nagelkerke		.34	.46	.35
Hoshmer Lemeshow		14.99 (8)*	7.95 (8)	9.80 (8)
WE Factors (Independent variables)	Usability	1.29 (.15)*	1.82 (.37)*	1.34 (.15)*
	Interactivity	.19 (.13)	-.56 (.28)	.18 (.13)
	Trust	.55 (.11)*	.96 (.21)*	.58 (.12)*
	Aesthetics	.47 (.12)*	.72 (.26)*	.47 (.12)*
	Mk Mix	.54 (.13)*	.71 (.25)*	.55 (.13)*
Main Motive (main reason for buying) (Independent variable)	To find better prices		-.15 (.40)	
Experience (Independent variable)	a: Years Internet usage			-.11 (.07)
	b: Online buyer			-.18 (.22)

According to the data above, as predicted, all WE elements have a positive effect on buying preferences. However, all of them, with exception of Interactivity factor, are significant influencers of the online buyers' preferences. So, the sub-hypotheses H1a, H1c, H1d and H1e are accepted. In contrast, the sub-hypothesis H1b is rejected. It means that the aesthetics factor is independent predictor of purchasing decision.

According to literature, Usability and Trust are the most influential web experience elements, while the Marketing Mix is a factor not very relevant in the online context. So, in this study, the H2 is rejected because in spite of Usability and Trust having relevant scores, Marketing Mix and Aesthetics have relevant scores too.

The most common motive for online shopping is "to find better prices". According to responses of participants, the "main motive" variable is not significant. It indicates that H3 is accepted because the motives (specifically, main motive) of online customers to buy through the Internet do not have an effect on the way the WE elements influence their online vendor preference.

After an isolated analysis on two experience variables we concluded that the "online buyer" variable is not significant (H4b is rejected), and the "years Internet usage" variable, with a confidence level of 92%, affects all factors except Interactivity. Therefore H4a (for all factors except Interactivity) can be accepted under these conditions. According to this study, the years of Internet usage affect the importance consumers attribute to some WE factors as influencers of their product and online vendor decisions, however, the fact that someone is an experienced online buyer does not affect the importance attributed.

6   Conclusions and Main Implications for e-Tailers

As major conclusion, in this work has been that Web Experience elements clearly influence online shopper preferences, in line with previous literature findings [e.g. 32; 33; 34; 29; 35]. The main influencers are the aspects related to Usability of web sites, as well as (although with less impact) the users' Trust towards website, Marketing Mix dimensions, and finally, the online store's Aesthetics elements. In this study one of the five WE elements (i.e. Interactivity) was not found to have any substantial influence on the choice of e-vendor.

Motives and familiarity of the Internet user with online purchasing do not seem to play any significant role on the online shopping process. In contrast, the years of web usage have significant effect on consumer preferences.

Based on [35], in this study we have tried to identify the possible existence of discrepancies between perceptions of e-consumers and what influences their buying decisions and final purchasing behavior. In this sense, despite the fact that a substantial percentage of participants perceive Interactivity as a relevant element affecting their choice of a virtual vendor, the actual buying behavior indicates that this is not the case.

In this study, navigational *click streams* were captured through register software. Nowadays, as future research line, we are analyzing these web registers in order to study each movement carried out by users and, consequently, analyze their different search ways and types of vendors chosen and rejected (through choice models).

We believe that an important implication for e-marketers involved in creating online stores is that the elements related to usability (e.g. easy navigation, simple order process, fast load of pages...), as well as aspects such us trust (i.e. guarantee and safety), marketing mix (e.g. attractive promotions, competitive prices, product assortment...), and also aesthetics (i.e. site's design and presentation) are important influencers of e-consumer preferences. Moreover, it is important to underline the fact that users' experience with the Internet has an effect on these elements, and turn, their preferences towards the online vendors.

References

1. Urban, G.: Customer Advocacy: Is it for you? Working Paper 175. MIT Sloan School of Management, Center for E-Business (2003), Available: http://ebusiness.mit.edu/research/papers/175_Urban_Trust.pdf
2. Novak, T., Hoffman, D., Yung, Y.: Measuring the Customer Experience in Online Environments: A Structural Modeling Approach. Marketing Science 19(1), 22–42 (2000)
3. Cappel, J.J., Myerscough, M.A.: World Wide Web uses for Electronic Commerce: towards a classification scheme (1996), Available: http://hsb.baylor.edu/ramsower/ais.ac.96/papers/aisor13.htm
4. Cockburn, C., Wilso, T.D.: Business use of the World Wide Web. International Journal of Information Management 16(2), 83–102 (1996)
5. Spiller, P., Lohse, G.L.: A classification of Internet retail stores. International Journal of Electronic Commerce 2, 29–56 (1997)
6. Jarvenpaa, S.L., Todd, P.A.: Consumer Reactions to Electronic Shopping on the World Wide Web. Journal of Electronic Commerce 1(2), 59–88 (1997)
7. Degeratu, D., Rangaswamy, A., Wu, J.: Consumer choice behavior in online and traditional supermarkets: The effects of brand name, price, and other search attributes. International Journal of Research in Marketing 17(1), 55–78 (2000)
8. Childers, T.L., Carr, C.L., Peck, J., Carson, S.: Hedonic and utilitarian motivations for retail shopping behavior. Journal of Retailing 77(4), 51–535 (2001)

9. Dahan, E., Hauser, J.: The virtual consumer: Communication, conceptualization and computation. Paper 104, MIT Sloan School of Management, Center for E-Business (2001), Available:
   http://ebusiness.mit.edu/research/papers/104%20EDahan,%20JHauser%20Virtual%20Cust omer.pdf
10. Eastin, M.: Diffusion of e-commerce: an analysis of the adoption of four e-commerce activities. Telematics and Informatics 19(3), 251–267 (2002)
11. Liu, X., Wei, K.K.: An empirical study of product differences in consumers' E-commerce adoption behavior. Electronic Commerce Research and Applications 2(3), 229–239 (2003)
12. Corbitt, B., Thanasankit, T., Yi, H.: Trust and e-commerce: a study of consumer perceptions. Electronic Commerce Research and Applications 2(3), 203–215 (2003)
13. van Schaik, P., Ling, J.: The effect of link colour on information retrieval in educational internet time. Computers in Human Behavior 19(5), 553–564 (2003)
14. Keen, W., Wetzels, M., de Ruyter, K., Feinberg, R.: E-tailers versus retailers: Which factors determine consumer preferences. Journal of Business Research 57, 685–695 (2004)
15. Nielsen NetRatings.: Coach, Marcus, N., Crew, J., maximize sales channel relationships with web customers, Press Release (2003), Available: http://www.nielsen-netratings.com/pr/pr_030114.pdf
16. Cai, S., Xu, Y.: Effects of outcome, process and shopping enjoyment on online consumer behavior. Electronic Commerce Research and Applications 5(4), 272–281 (2006)
17. Tan, W.G., Wei, K.K.: An empirical study of Web browsing behaviour: Towards an effective Website design. Electronic Commerce Research and Applications 5(4), 261–271 (2006)
18. Eroglu, S.A., Machleit, K.A., Davis, L.M.: Atmospheric qualities of online retailing: a conceptual model and implications. Journal of Business Research 54, 177–184 (2001)
19. Eroglu, S.A., Machleit, K.A., Davis, L.M.: Empirical testing of a model of online store atmospherics and shopper responses. Psychology and Marketing 2(2), 139–150 (2003)
20. Vrechopoulos, A.P., O'Keefe, R.M., Doukidis, G.I.: Virtual Store Atmosphere in Internet Retailing. In: Proc. 13th International Bled Electronic Commerce Conference, Bled Slovenia, pp. 19–21 (2000)
21. Vrechopoulos, A.P., Siomkos, G.J.: Virtual store atmosphere in non-store retailing. Journal of Internet Marketing 3(1), 22–38 (2002)
22. Vrechopoulos, A.P.: An emerging store layout for Internet grocery retailing. In: Proc. 3rd ECR Europe Scientific Competition, Barcelona, Spain (2002)
23. Vrechopoulos, A.P.: Developing alternative store layouts for Internet retailing. In: Doukidis, G., Vrechopoulos, A.P., Consumer Driven Electronic Transformation: Apply New Technologies to Enthuse Consumers, Springer, Heidelberg (2004)
24. Dailey, L.: Navigational web atmospherics. Explaining the influence of restrictive navigation cues. Journal of Business Research 5754, 1–9 (2004)
25. Lorenzo, C., Gómez, M.A., Mollá, A.: Impact of hedonic attributes on consumer responses: Animation of products in the virtual world. In: Proc. 13th EIRASS Conference, Budapest, Hungary, pp. 9–12 (2006)
26. Lindgaard, G., Fernandes, G., Dudek, C., Brown, J.: Attention web designers: You have 50 milliseconds to make a good first impression. Behaviour & Information Technology 25(2), 115–126 (2006)
27. Tractinsky, N., Cokhavi, A., Kirschenbaum, M., Sharfi, T.: Evaluating the consistency of immediate aesthetic perceptions of web pages. International Journal of Human-Computer Studies 64(11), 1071–1083 (2006)

28. Watchfire Whitepaper Series: Bad Things Shouldn't Happen to Good Web Sites, Best Practices for Managing the Web Experience (2000), Available: http://www.watchfire.com/resources/search-and-ye-shall-find.pdf
29. Constantinides, E.: Influencing the Online Consumer's Behavior: The Web Experience. Journal of Internet Research: Electronic Networking Applications and Policy 14(2), 111–126 (2004)
30. Hoffman, D., Novak, T.: Marketing in Hypermedia Computer-Mediated Environments: Conceptual Foundations. Journal of Marketing 60, 50–68 (1996)
31. Csikszentmihalyi, M.: Flow: The Psychology of Optimal Experience. HarperPerennial, New York (1990)
32. Koufaris, M.: Applying the technology acceptance model and flow theory to online consumer behavior. Information Systems Research 13(2), 205–223 (2002)
33. Pace, S.: A grounded theory of the flow experiences of Web users. International Journal of Human-Computer Studies 60, 327–363 (2004)
34. Kuniavsky, M.: Observing the user experience. A practitioner's guide to user research, 1st edn. The Morgan Kaufmann Series in Interactive Technologies (2003)
35. Constantinides, E., Geurts, P.: The impact of web Experience on virtual buying behavior: An empirical study. Journal of Customer Behavior 5, 307–336 (2006)
36. O'Keefe, R.M., McEachern, T.: Web-based Customer Decision Support Systems. Communications of the ACM 41, 71–78 (1998)
37. Davis, F.D.: Perceived usefulness, perceived ease of use, and user acceptance of information technology. MIS Quarterly 13(3), 319–340 (1989)
38. Yoh, E., Damhorst, M.L., Sapp, S., Laczniac, R.: Consumer adoption of the Internet: The case of apparel shopping. Psychology & Marketing 20(12), 1095–1118 (2003)
39. Nielsen, J.: Usability 101 (2003), Available: http://www.useit.com
40. Flavián, C., Guinalíu, M., Gurrea, R.: Análisis empírico de la influencia ejercida por la usabilidad percibida, la satisfacción y la confianza del consumidor sobre la lealtad a un sitio web. In: Proc. XVI Encuentro de Profesores Universitarios de Marketing, Alicante (Spain), pp. 209–226 (2004)
41. Flavián, C., Guinalíu, M., Gurrea, R.: The role placed by perceived usability, satisfaction and consumer trust on website loyalty, Information and Management (2005), Available: http://www.elsevier.com/locate/dsw
42. O'Cass, A., Fenech, T.: Web retailing adoption: Exploring the nature of Internet users web retailing behaviour. Journal of Retailing and Consumer Services 10, 81–94 (2003)
43. McCarthy, E.J.: Basic Marketing: A managerial approach, 2nd edn., Irwin, R. D. (1964)
44. Goldsmith, R.E.: The Personalized Marketplace: Beyond the 4Ps. Marketing Intelligence & Planning 17(4), 178–185 (1999)
45. Jobber, D.: Principles and Practice of Marketing. McGraw-Hill, London (2001)
46. Kotler, P.: Marketing Management, 11th edn. Prentice-Hall International Editions, Englewood Cliffs,NJ (2003)
47. Constantinides, E.: The 4s web-marketing mix model, e-commerce research and applications. Elsevier Science 1(1), 57–76 (2002)

A Framework for Defining Fashion Effect in Electronic Commerce Environments

Dorin Militaru

Amiens School of Management, 18 place Saint Michel - 80 000 Amiens,
FRANCE GRID, 30 Avenue du Président Wilson, 94235 Cachan, France
dorin.militaru@supco-amiens.fr

Abstract. The main purpose of the study is to investigate how different dimensions of consumer perception and consumer attitude may affect behaviour in electronic commerce environments. Consumer behaviour, its antecedents and outcomes, and the relationship between consumer perception, consumer attitudes and fashion effect have been analyzed at a customer level. The results indicated that fashion effect affect consumer's behaviour in electronic commerce environments. The theoretical concepts that form the foundation of the paper appear to have a significant application to consumer marketing, but more studies are needed. The main contribution of this paper is the completion of an exhaustive analysis of the fashion effect and its antecedents. On one hand, it is a pioneer in the study of the influence of fashion effect on electronic commerce and, on other hand, it confirms results from other researches in the traditional commerce environments.

Keywords: Consumers, Europe, Marketing, Consumer behaviour, Internet shopping, Electronic commerce, Fashion effect.

JEL classification: C12, D11, D12, M20, M31.

1 Introduction

The impact of Internet on consumer behaviour and marketing practices is now a major and challenging area of research. Internet offers many choices of products, services and content. But the multitude of choices has altered the manner in which customers choose and buy products and services. Among the several new situations or possibilities raised by the emergence of Internet is rapid and more expansive communications in a virtual world without face-to-face interaction. Hypothetically, such an environment should facilitate the identification of needs or desires which stimulate consumers first, to search for products or services, and afterwards to purchase them. Indeed, this electronic environment is starting to modify the more traditional perceptions and attitudes that take place in traditional markets [1].

The role of the customers in this competitive environment becomes ever more important since they convey ideas, impressions or feelings on website stores, concentrated in "customers' opinions" or on costumer forums. In addition, buyers are, in someway, at the interface between retailers and consumers, and their presence is highly influential in maintaining the relationship between them. Thus, electronic

G. Psaila and R. Wagner (Eds.): EC-Web 2007, LNCS 4655, pp. 201–211, 2007.
© Springer-Verlag Berlin Heidelberg 2007

buying, while ostensibly economic in function, demands acute and detailed perceptional knowledge, and finely honed social skills. Group influences, such as those associated with belonging to or identifying with a social class, culture or subculture, or reference groups, such as family and other social groups became even more influential in electronic environments [2]. However, the specific details regarding consumer perceptions, attitudes on Internet-based buying situations have only received limited attention. This study attempts to close this gap. Through an extension of [3] study, the main objectives of this paper are to investigate how different dimensions of consumer perceptions and attitudes affect the buying decision process and to test the possibility of generalizing the results with a European sample.

2   Theoretical Framework and Main Hypotheses

The consumer buying behaviour is influenced by cultural, social, personal and psychological factors. In general, an individual's buying choices are influenced by the psychological factors of motivation, perception, learning, beliefs, and attitudes [4]. While [5] suggests that more psychological factors should be integrated in order to describe decision process in electronic environments, our objective is to focus exclusively on perceptions and attitudes and to assess their value in creating an effect designate as "fashion effect".

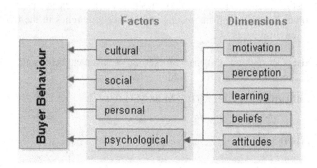

Fig. 1. Factors Influencing Buyer Behavior

2.1   Factors Influencing Buyer Behaviour - Relevant Literature

Perceptions. [6] define perception as the process by which an individual selects, organizes, and interprets information inputs to create a meaningful picture of the world. Future studies find that perception depends not only on physical stimuli, but also on the stimuli's relation to the situation and on conditions within the individual [7], [8]. It is assumed that a motivated person is prepared to take decisions. His perceiving of a certain situation influences his decisions. Others studies have shown [9] that perceptual processes are characterized by attention, distortion, and retention. These findings allow us to explain how the same person can have different perceptions of the same product. However no study has investigated how a perceptual process operates in electronic environments.

On Internet for example there are various stimuli such as advertisements which are filtered by a selective attention. People take into account stimuli that relate to a current need or stimuli that they anticipate or expect. Furthermore, these noticed stimuli are modified and interpreted in a way that fits theirs presumptions. Finally, individuals are likely to remember the perceived positive value for the products they already know rather than for new products.

Attitudes. [10] claims that people acquire attitudes by doing an action and learning the outcomes. Moreover buying behaviour is influenced by these personal attitudes. Beliefs are related with attitudes and they may be based on knowledge, opinion, or faith [11]. More important, these beliefs make up product, brand or situation images. Individuals act according to these images and we think that beliefs and attitudes play a role in the formation of the fashion effect. [12] defined an attitude as a person's positive or negative evaluations, emotional feelings, and action tendencies toward some object or idea, and which is stable over time. Attitudes put individuals into a state of mind of liking or disliking an object, idea or situation moving toward or away from it. As result, for similar objects, people behave in a fairly consistent way. In theory, attitudes economize on energy and thought, and as a consequence, they are very difficult to change. Despite this idea, we think than an individual adoption process of a group attitude may require fewer adjustments in other attitudes, like for example in the case of opinion leaders. Studies have found also that buyers often hold distinct attitudes about brands or products based on their country of origin [13]. Research on individual factors suggests that each buyer carries personal motivations, perceptions or preferences and exhibits different buying styles.

Social Factors. In order to explain how an individual adopts other people values and, by doing this, takes the same actions without necessarily borrowing the same decision process, we explore the influence of social factors. Reference groups, family or even the more impersonal mass media have a direct influence on a person's perceptions and attitudes [14], [15]. These social factors expose individuals to new behaviours and lifestyles, influence attitudes and perceptions, and more important, create pressures for conformity that may affect product and brand choices.

Principal reference groups are family, friends, neighbours, and colleagues, with whom individuals interact continuously and informally [16]. Individuals are also influenced by groups to which they do not belong, which is the case with electronic communities. For the process of values adoption we focus our attention on those whose values or behaviour an individual wants to adopt, called aspirational group. We think that aspirational groups in electronic environment are created by mass media and electronic media at the same time as the evaluation stage in individual buying decision process.

2.2 Defining Fashion Effect

The word "fashion" is defined as a currently popular style of clothing, a behaviour, or a manner of doing something[1]. In this paper we regard fashion as a behaviour or a manner of doing something. Moreover, we define fashion effect as an appropriation

---

[1] Concise Oxford Dictionary, 2003.

process of group attitudes which induces perceptions construction and affects decision-making process over product choices subset. It is related with a given situation or a given product and maybe a way of life. Measuring this effect could raise some technical problems. However we think that an appropriate measure must be a continuous construct.

The fashion effect, such we define it, is different from two other types of concepts namely fashion-oriented impulse buying and fashion clothing involvement. Moreover, we look at the current state of research in consumer behaviour that can provide the theoretical foundations underlying fashion effect, which for example, is not concerned as such with fashion products but keeps in mind the preoccupations of consumers to behave like others consumers. For clothing, research on fashion-oriented impulse buying [17] refers to a person's awareness or perception of fashionability attributed to an innovative design or style. Some authors suggest that it is a helpful metric for explaining consumer behaviour and segmenting consumer markets [18], [19], [20]. This type of impulse buying occurs when consumers see a new fashion product and buy it because they are motivated by the suggestion to buy new products [21]. Previous research has sought to understand the ways in which consumers become involved with products and to understand the effect involvement has on various purchasing and consuming behaviours [22], [23]. Early research into impulse buying behaviour [21] concentrated on the typology of impulse buying. According to this author, impulse buying was classified as four types: planned impulse buying; reminded impulse buying; fashion-oriented impulse buying; and pure impulse buying.

2.3 Antecedents of Fashion Effect

Positive emotions. Many studies have revealed that emotion strongly influences consumer's decision making [3], [24]. Typically, emotion has a positive and a negative dimension [25]. Several qualitative studies reported that in positive emotional states consumers tend to have reduced decision complexity and shorter decision times [5]. Moreover, consumers with positive emotion exhibited greater impulse buying because of feelings of being unconstrained, a desire to reward themselves, and higher energy levels [26]. [24] found consumer's positive emotion was associated with the urge to buy impulsively. Because impulse buyers exhibit greater positive feelings - like for example pleasure, excitement, and joy - they often overspend when shopping [27]. Therefore, consumer emotion can be an important factor for determining fashion effect in an electronic environment.

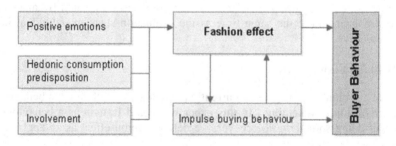

Fig. 2. Factors influencing fashion effect

Hedonic predisposition. Hedonic consumption is a behavioural aspect related to sensorial, fantasy, and emotional consumption. Individuals experience perceived-benefits such as fun using the product and esthetical appeal [28]. [29] identifies that bargaining and haggling are two shopping experiences associated with shopping enjoyment. This suggests that the purchasing situation may be more important than product acquisition. Impulse buying plays an important role in fulfilling hedonic desires associated with hedonic consumption [3], [30]. Others authors have found that hedonic' desires or non-economic reasons, such as fun, fantasy, and social or emotional gratification motivate impulse buying [3], [31]. Since the shopping experience goal is to satisfy hedonic needs, consumers involved in this process can help us to explore the attitudes appropriation process and by doing this the fashion effect.

Involvement. Involvement is used principally to predict behavioural variables related to apparel products such as product involvement, buying behaviour, and consumer characteristics [22], [32]. Consumer researchers have shown considerable interest in values because they are argued to be an important influence on behaviour. Socio-demographic characteristics have been identified as antecedents to involvement in various possessions, including both gender and age. [33] for example, argue that women are more involved in fashion and [34] argues that men more involved in cars. [5] found apparel impulse buying was distinguished from reasonable unplanned buying that was based on emotional preference or objective evaluation rather than rational evaluation.

3 Main Hypotheses

With regard to buying behaviour theory, there are two potential methods for measuring fashion effect. One is to measure individuals' psychological factors in terms of how a consumer actually perceives and regards a product or a situation. The other is to measure psychological factors in terms of preferences over a products / situations subset. The first approach is related to the objective psychological factors of an individual held in long-term memory, while the second approach is inferred from an individual's subjective self-report on why they prefer an item to another. A fundamental question that arises in attempting to understand an appropriation process is what

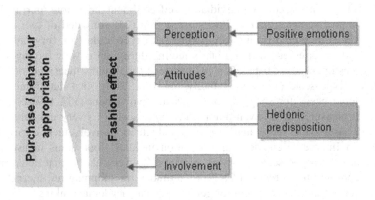

Fig. 3. Causal relationships

consumers know about their perceptions and attitudes, and what characteristics cause some consumers to borrow a decision process from another individual.

The issue raised herein is that fashion effect has a number of key aspects, namely positive emotions, hedonic predisposition, attitudes and perceptions. In a general sense we would assume that fashion effect will result in consumer's increased mimetic behaviour.

In this regard, the perspective taken here is that the degree of influence caused by fashion effect in decision buying process should be affected by the consumers' propensity to adopt other people attitudes and perceptions. It is hypothesised that:

H1: Positive emotions have a significant positive effect on group attitudes appropriation.
H2: Positive emotions have a direct influence on consumer perceptions.
H3: Hedonic predisposition has a significant positive effect on behaviour appropriation.
H4: Consumer involvement has a significant positive effect on behaviour appropriation.
H5: Consumer perceptions have a significant positive influence on behaviour appropriation.
H6: Consumer attitudes have a significant positive influence on behaviour appropriation.
H7: Consumer attitudes are more important than involvement in behaviour appropriation.

4. Research Method

4.1 Measurement of Variables

The research method has been designed as to completely fulfil the objectives of this study and bring a comprehensive and exhaustive answer to each issue raised herein. The research model depicted in Figure 3 was developed to examine buying behaviour appropriation process. It illustrates the causal relationships among six variables (perceptions, attitudes, positive emotion, hedonic predisposition, involvement, and behaviour appropriation) in an Internet purchasing situation. A self-administered survey included these five variables was developed and administered via email. To measure perceptions [35], [36] we consider two items (product, buying situation) measured on a seven-point Likert-type scale (1 - very unlikely, 7 - very likely) that assessed an individual's feelings interpreted in the light of experience during the last purchase in an Internet situation. Attitudes [37], [38] consisted of the same two items (product, buying situation) measured on a seven-point Likert-type scale (1 - very unlikely, 7 - very likely) that assessed an individual's self-confident behaviour during the last purchase in an Internet situation. Positive emotion [24] consisted of two items (excited, satisfied) measured on a seven-point Likert-type scale (1 - very unlikely, 7 - very likely) that assessed an individual's feeling during an on-line buying situation. Hedonic consumption predisposition [3] included three items measured on a seven-point Likert-type scale (1 - very unlikely, 7 - very likely) that determined respondents' hedonic needs for shopping on Internet. Involvement [17], [39] consisted of three items measured on a seven-point Likert-type scale (1 - very unlikely, 7 - very likely) that assessed an individual's product and situation involvement during the last purchase on Internet. Behaviour appropriation included four items measured on a seven-point Likert-type scale (1 - very unlikely, 7 - very likely) that determined respondent appropriation process of a group attitudes for shopping on Internet. Demographic information was collected for gender, age and academic ranking.

4.2 Sampling Procedures

The study used individuals with an e-mail address. Respondents came from a random sample of consumers drawn from a database containing 10 000 names from residents living Europe. In total 2 000 names were randomly drawn from the list provided. The study design excluded those individuals that haven't made any purchases over the Internet. The goal was to obtain 400 completed questionnaires. The self-administered Internet questionnaire contained 20 items and lasted an average of 12 minutes.

4.3 Data Collection Procedures and Data Analysis

Obtaining complete responses to a questionnaire remains problematic for any research method, but particularly for self-administered Internet studies. In a first time, each selected participant received an e-mail seeking their participation in the study. In a

Table 1. Exploratory factor analysis of construct items

Factor labels and statements	Factor loading	Reliability	Variance extracted
Perceptions		0.80	0.68
On Internet …			
I give more attention to products reviews	0.86		
I am more likely to notice information about web-sites stores	0.70		
Attitudes		0.81	0.70
On Internet I take into account …			
Evaluations about product that came from other people	0.82		
Opinions on websites that came from other people	0.73		
Positive emotion		0.85	0.79
When buying on-line …			
I am excited	0.95		
I am satisfied	0.81		
Hedonic predisposition		0.91	0.81
When buying on-line …			
I want to satisfy my sense of curiosity	0.90		
I want to be offered new experiences	0.95		
I want to feel like I'm exploring new worlds	0.83		
Involvement		0.82	0.65
For me personally buying on-line is important	0.86		
I am interested in products for sale on Internet	0.80		
I am interested in shopping at on-line boutique	0.72		
Behaviour appropriation		0.90	0.81
When buying on-line …			
I take into account the advice from other buyers	0.94		
I chose websites with goods recommendations	0.85		
The others costumers help me to chose products	0.82		
I sped less time because I learn from others costumers' experiences	0.92		

Note: Means are reported on a seven-point scale (1 - low ; 7 - high)
All constructs were operationalised as perceived constructs

Table 2. Partial least squares results for the theoretical model

Predicted variables	Predictor variables	Hypothesis	Path weight	Variance due to path	R^2	Critical ratio
Attitudes appropriation	Positive emotions	H1	0.524	0.27		3.98
Consumer perceptions	Positive emotions	H2	0.336	0.16		6.82
Behaviour appropriation	Hedonic predisposition	H3	0.178	0.06	0.31	3.55
	Consumer involvement	H4	0.629	0.49		29.43
	Consumer perceptions	H5	0.605	0.32	0.42	9.69
	Consumer attitudes	H6	0.578	0.29	0.25	8.11
AVA					0.33	

Note: AVA = average variance accounted

second time, we send a reminder e-mail message. A total of 109 questionnaires were rejected due to incomplete answers. Useful data were obtained from 451 Europeans respondents. This makes a final total useful response rate of 22.5 percent. The exact structure of the sample was 14.8 per cent under 30 years of age, 31-40 age group constituted 46.2 per cent of the respondents, 41-50 age group were 26.4 per cent of respondents and the over 50 years group constituted 12.6 per cent of the sample. Over 54.2 per cent of respondents were male and 45.8 per cent were female. Before testing the hypotheses the properties of the scales were examined. The data were initially examined for dispersion and central tendency via means, standard deviation and skew and kurtosis, with the analysis indicating no anomalies in the data.

Measurement model and structural model was tested using LISREL 8.5. Following below, items within each construct were then computed into composite variables to test the models. Forming composites is a generally accepted approach in consumer behaviour to test hypotheses [39]. Concerning discriminate validity, the results indicated that all reliability estimates (Cronbach's alpha) were greater than their correlation. The estimated measurement model consisted of two items for perceptions, two items for attitudes, two items for positive emotion, three items for hedonic predisposition, three items for involvement, and four items for behaviour appropriation process. Overall, the coefficients of factor loading on the latent constructs ranged from 0.70 to 0.95, $p < 0.001$. Reliabilities of the latent variables ranged from 0.80 to 0.91 and confirmed the measurement model was valid and reliable.

Descriptive analysis revealed above average mean scores for each research construct: perceptions M = 4.32, attitudes M = 4.54, positive emotion M = 4.21, hedonic predisposition M = 4.38, involvement M = 4.73 and behaviour appropriation M = 4.58.

For testing the hypotheses, consideration was given to appropriate analytical techniques. Given the formulation of the hypotheses it was decided to use partial least squares method which is a variance based general regression technique for estimating path models involving latent constructs simultaneously [40], [41]. A methodical analysis of a number of fit indices was used to assess the predictive relevance of the hypotheses. It includes R2 and average variance accounted for regression weights and loadings [41]. These indices provide evidence for the existence of the relationships rather than definitive statistical tests, which may be contrary to the philosophy of soft modelling [40].

5 Results

Results show that the average variance accounted for in the endogenous variable by the exogenous variables was 0.33. According to [40], for all of the predicted variables the specific values for R2 were greater than the recommended 0.10. Consequently, examination of the significance of the paths associated with these variables was also undertaken. We use the absolute value of the product of the path coefficient and corresponding correlation coefficient to evaluate the significance of the individual paths. Because paths are estimates of the standardized regression weights this produces an index of the variance in an endogenous variable explained by that particular path. In this case 0.015 of the variance is recommended as the cut off point. In Table 2, all the paths exceed this criterion and the magnitude of the paths was appropriate. Likewise critical ratios are greater than 1.96.

The data analysis suggests that positive emotions have a significant impact on attitudes appropriation, where in effect, stronger positive emotions tendency brings individuals to adopt easily the group attitudes. Also positive emotions have a direct influence on consumer perceptions. As well, hedonic predisposition, consumer involvement, consumer perceptions and consumer attitudes have a significant positive influence on behaviour appropriation. Therefore, hypotheses H1-H6 are supported. We notice that hedonic predisposition path coefficient indicates a smaller effect of this exogenous variable then the other variables tested. Concerning hypothesis H7, the data suggests that consumer involvement are more important than consumer attitudes in behaviour appropriation process. Therefore, this hypothesis is not supported.

6 Conclusions

Contrary to most expectations, it appears that, for a consumer, it may become more difficult to identify his or her own needs and preferences on electronic environments, and the ways to satisfy them [2]. In addition, internet and related technologies do allow for rapid and useful communication. The research framework appears to be an important support in understanding consumers and their purchasing and consumption related behaviour on Internet. This extends into what cognitive and behavioural process cause some consumers to perceive product and situations in different ways in online buying situations. Such an approach can only enhance the effective utilization of consumer characteristics at a theoretical and practical level in understanding consumer behaviour related to electronic commerce. This is important because electronic commerce has both important economic and social significance in Europeans societies. The most important contribution of the present study was to define a new effect, called fashion effect. The key idea was that buyers on Internet operate with a keen sense of the relationship between their areas of experiences. Buyers build up a picture of markets through their encounters with them. Much of this interaction is implicit and depends upon the information acquisition process. Externally, buyers also build up a picture of market through the relationship between the websites stores and their customers. Moreover, there are implications that attitudes appropriation, perceptions and positive emotion are important predictors of consumers' on-line buying. From a hedonic perspective, positive emotion increased fashion effect, whereas hedonic consumption did not relate directly to buying behaviour appropriation. This finding suggests that on Internet, decision-making process in the classical theory of consumer behaviour may change. However, further research on this field is needed.

References

1. Rosen, E.: The Anatomy of Buzz. Doubleday, New York, NY (2000)
2. Vézina, R., Militaru, D.: Collaborative Filtering: Theoretical Positions and a Research Agenda in Marketing. International Journal of Technology Management 28(1), 31–45 (2004)
3. Hausman, A.: A multi-method investigation of consumer motivations in impulse buying behavior. Journal of Consumer Marketing 17(15), 403–419 (2000)
4. Kotler, P.: Marketing Management Millenium Edition, 10th edn. Prentice-Hall, Englewood Cliffs (2000)
5. Isen, A.M.: The influence of positive affect on decision-making and cognitive organization. Advances in Consumer Research 11, 534–537 (1984)
6. Berelson, B., Steiner, G.A.: Human Behavior: An Inventory of Scientific Findings, pp. 493–525. Harcourt Brace & World, New York (1964)
7. Herzberg, F.: Work and the Nature of Man, Cleveland: World Publishing Company (1966)
8. Thierry, H., Koopman-Iwerna, A.M.: Motivation and Satisfaction. In: Drenth, P.J. (ed.) Handbook of Work and Organizational Psychology, pp. 141–142. John Wiley, New York (1984)
9. Kassarjian, H.H., Sheffet, M.J.: Personality and Consumer Behavior: An Update, Perspectives in Consumer Behavior, 160-180 (1981)
10. Fishbein, M.: Attitudes and Prediction of Behavior. In: Fishbein, M. (ed.) Readings in Attitude Theory and Measurement, pp. 477–492. John Wiley, New York (1967)
11. Sheth, J.N.: An Investigation of Relationships among Evaluative Beliefs, Affect, Behavioral Intention, and Behavior. In: Farley, J.U., Howard, J.A., Ring, L.W. (eds.) Consumer Behavior: Theory and Application, Boston, pp. 89–114 (1974)
12. Krech, D., Crutchfield, R., Ballachey, E.: Individual in Society, ch. 2. McGraw-Hill, New York (1962)
13. Phau, I., Suntornnond, V.: Dimensions of consumer knowledge and its impacts on country of origin effects among Australian consumers: a case of fast-consuming product. Australia Journal of Consumer Marketing 23(1), 34–42 (2006)
14. Lepisto, L.: A Life Span Perspective of Consumer Behavior. Advances in Consumer Research 12, 47 (1985)
15. Sheehy, G.: New Passages: Mapping Your Life Across Time. Random House (1995)
16. Wortzel, L.H.: Marital Roles and Typologies as Predictors of Purchase Decision Making for Everyday Household Products: Suggestions for Research. Advances in Consumer Research 7, 212–215 (1989)

17. Jones, M.A., Reynolds, K.E., Weun, S., Beatty, S.E.: The-product-specific nature of impulse buying tendency. Journal of Business Research 56(7), 505–511 (2003)
18. Kapferer, J.-N., Laurent, G.: Measuring consumer involvement profile. Journal of Marketing 22(1), 41–53 (1985)
19. Kim, H.-S.: Consumer profiles of apparel product involvement and values. Journal of Fashion Marketing and Management 9(2), 207–220 (2005)
20. Martin, C.L.: Relationship marketing: a high-involvement product attribute approach. Journal of Product & Brand Management 7(1), 6–26 (1998)
21. Han, Y.K., Morgan, G.A., Kotsiopulos, A., Kang-Park, J.: Impulse buying behaviour of apparel purchasers. Clothing and Textiles Research Journal 9(3), 15–21 (1991)
22. Browne, B.A., Kaldenberg, D.O.: Conceptualizing self-monitoring: links to materialism and product involvement. Journal of Consumer Marketing 14(1), 31–44 (1997)
23. Traylor, M.B., Joseph, B.W.: Measuring consumer involvement in products. Psychology and Marketing 1(2), 65–77 (1984)
24. Beatty, S.E., Ferrell, E.M.: Impulse buying: modelling its precursors. Journal of Retailing 74(2), 169–191 (1998)
25. Watson, D., Tellegen, A.: Toward a consensus structure of mood. Psychological Bulletin 98(2), 219–335 (1985)
26. Rook, D.W., Gardner, M.P.: In the mood: impulse buying's affective antecedents. Research in Consumer Behavior 6, 1–26 (1993)
27. Donovan, R.J., Rossiter, J.R.: Store atmosphere: an environmental psychology approach. Journal of Retailing 58(1), 34–57 (1982)
28. Hirschman, E.C., Holbrook, M.B.: The experiential aspects of consumption: consumer fantasies, feelings, and fun. Journal of Consumer Research 9(2), 132–140 (1982)
29. Sherry, J.F.: A sociocultural analysis of a Midwestern American flea market. Journal of Consumer Research 17(1), 13–30 (1990)
30. Rook, D.: The buying impulse. Journal of Consumer Research 14(2), 189–199 (1987)
31. Piron, F.: Defining impulse purchasing. Advances in Consumer Research 18, 509–513 (1991)
32. Flynn, L.R., Goldsmith, R.E.: A causal model of consumer involvement: replication and critique. Journal of Social Behavior and Personality 8(6), 129–142 (1993)
33. Tigert, D.J., King, C.W., Ring, L.R.: Fashion involvement: a cross-cultural analysis. Advances in Consumer Research 17, 17–21 (1980)
34. Bloch, P.H.: Involvement beyond the purchase process: conceptual issues and empirical investigation. Advances in Consumer Research 9, 413–430 (1982)
35. Babin, B.J., Babin, L.: Seeking something different? A model of schema typicality, consumer affect, purchase intentions and perceived shopping value. Journal of Business Research 54(2), 89–96 (2001)
36. Day, G.: Buyer Attitudes and Brand Choice Behavior. The Free Press, New York (1970)
37. Harrell, G.: Industrial product class involvement, confidence in beliefs and attitude intent relationships, In: Attitude Research Conference 8th Las Vegas 1979: Attitude Research Plays for High Stakes, American Marketing Association, Chicago, IL, pp. 133-147 (1979)
38. O'Cass, A.: An assessment of consumers' product, purchase decision, advertising and consumption involvement in fashion clothing. Journal of Economic Psychology 21(5), 545–576 (2000)
39. Falk, F.R., Miller, N.B.: A Primer for Soft Modeling. University of Akron Press (1992)
40. O'Cass, A.: Consumer self-monitoring, materialism and involvement in fashion clothing. Australasian Marketing Journal 9(1), 46–60 (2001)

DRLinda: A Distributed Message Broker for Collaborative Interactions Among Business Processes*

J. Fabra, P. Álvarez, and J. Ezpeleta

Instituto de Investigación en Ingeniería de Aragón (I3A)
Department of Computer Science and Systems Engineering, University of Zaragoza,
María de Luna 3, E-50018 Zaragoza (Spain)
{jfabra,alvaper,ezpeleta}@unizar.es

Abstract. Recently, coordination middleware systems have evolved in order to describe coordination protocols in business process scenarios. This evolution proposes the use of three main components, being one of them a message broker to handle collaborative interactions among business processes. In a previous work, we proposed a framework for coordination in open BPM systems which used a centralised Linda-based implementation of a message broker. The use of a centralised implementation leads to some common problems which a distributed model tries to solve in an efficient manner. In this paper, we present *DRLinda*, a distributed and dynamic implementation of the message broker based on the RLinda model, which improves and extends the RLinda's features and can be configured at runtime, being suitable for more complex and highly-dynamic business process scenarios. The performance of the proposed implementation is empirically evaluated on a cluster computing environment.

Keywords: EC Infrastructure and Basic Technologies, Business Process Aspects, Service Coordination, Linda, Petri Nets.

1 Introduction

The core functionality of Web processes middleware tries to wrap existing business logics making them accessible as Web services. Recently, well-known standardisation initiatives have proposed some high-level declarative languages for the description of coordination protocols and the implementation of coordination middleware. The evolution of such middleware requires the addition of some functionalities already present in traditional enterprise application integration platforms. These middleware systems contain three main components [1,2]: 1) a workflow engine for the definition and execution of the business logic, 2) a message broker to route messages among participants and 3) the necessary infrastructure to ensure the correctness and consistency of interactions (correct integration and execution of both, horizontal and business protocols).

* This work has been supported by the research project PIP086/2005, granted by the Government of Aragón and the project TIN2006-13301, granted by the Spanish ministry of Education and Science.

G. Psaila and R. Wagner (Eds.): EC-Web 2007, LNCS 4655, pp. 212–221, 2007.

A proposal of Web service middleware for the definition and execution of (complex) web processes which can interact using (complex) choreographies and the conversational approach was presented in [3,4]. The base formalism used in the proposal was the *Reference nets* paradigm [5] which is a subclass of the family of *Nets-within-Nets* [6]. As shown in the paper, the use of the same formalism for composition and coordination purposes helps with the simplification of the development process.

The framework proposed in [3] represents an evolution of the one described in [4] and contains three main components. Firstly, the *composition component* which integrates the workflow engine executing the business logic of Web processes. Any standard language (such as BPEL4WS) could be used for the definition and description of the considered workflows. Secondly, the *conversation component* which handles conversations among processes involved in a started business transaction and implements the logic of interactions. Finally, the *message broker* which separates the logics of the message interchanges from the concrete way in which a message must be delivered or received. Besides, depending on the destination, different transport protocols (RPC, HTTP, SMTP, etc..) can be used.

A coordination system based on the Linda paradigm was used as an intermediate language for the definition of the interactions among the involved Web processes and participants. This system, called RLinda [3,7] (Renew-based Linda), acts as a message repository and provides with an abstract and technology-independent definition of the inter-processes interactions. The use of a centralised implementation leads to some common problems which a distributed model tries to solve in an efficient manner. In this work, we present our first approach towards a distributed implementation of the Linda coordination system based on the RLinda model, called DRLinda – Distributed RLinda –, which improves and extends the RLinda's features and can be configured at run-time, being suitable for complex and highly-dynamic business process scenarios.

The paper is organised as follows. Section 2 introduces the Linda coordination paradigm, the problems derived from a centralised implementation and presents some distributed implementations. Section 3 describes our distributed implementation of the Linda coordination system, DRLinda. Section 4 presents the empirical performance measures obtained, showing also the speed-up when a set of distributed RLinda nodes are coordinated. Finally, some conclusions are presented.

2   Towards a Distributed Linda Implementation

The Linda coordination model, originally proposed by Carriero and Gelernter [8,9], was based on the principle of generative communication, where a global tuple space is accessed by a set of processes with a limited set of operations: an *out* operation to put tuples into the space, an *in* operation in order to perform a destructive read from the tuple space and a *rd* operation to achieve a non-destructive read. The reading operations can receive a pattern as a parameter in order to withdraw a matching tuple from the space, blocking the caller until the tuple is returned.

Although Linda's conceptual model represents an easy coordination model, its implementation is more complex. The main problem appears in the physical representation of tuples and in the insertion, deletion and matching algorithms. The design and implementation of these topics will affect in a decisive manner the system performance. A centralised implementation suffers from some lacks derived from the existence of a single tuple repository, as being non fault tolerant, making a bad load balance or making difficult to deal with security aspects.

A distributed implementation can help in dealing properly with these aspects, although it requires to take care of additional considerations related to the tuple definition and operations. On the one hand, Linda tuple's semantics must be kept, what suggests the existence of an efficient API design and construction. The aim of using an API for tuples and templates is to hide implementation and storage details, opening the possibility of using different implementations and mechanisms for distributed versions. On the other hand, tuples have an unlimited persistence in the original Linda model, which is a very hard requirement for distributed environments. The distributed nature of the systems we are considering implies some considerations for the tuple space operations. Firstly, the tuple insertion. Location policies must be designed with a minimum impact on the overhead that the use of a load balancing algorithm implies. The scalability of the whole system is directly affected by the choice of the algorithm. Secondly, tuple removal / deletion operations need to consider that tuples removed from the space must be inaccessible for the rest of system nodes. Finally, the implementation of a tuple search algorithm for a distributed Linda coordination system must keep the original operation semantics.

The Linda coordination language is orthogonal to the rest of system components, which has produced lots of different centralised implementations. Due to the scope of this work, we are only going to briefly describe some existing distributed implementations which try to consider and solve some of the previous points. An implementation oriented to mobile adaptation scenarios was presented in [10]. This work suggests the separation of the communication among processes into several levels. A tuple space can be created on each one of these levels, creating thus a hierarchical organisation of system nodes. This proposal points towards scalability criteria, but it lacks of a defined specification of levels and spaces. A similar approach is adopted in [11], where a logical separation is proposed in the hierarchical structure. It is possible to differentiate between Remote Tuple Spaces (RTS) and Local Tuple Spaces (LTS). RTS are stored in tuple space servers (TSS), and LTS interacts with RTS using a manager (LTSM). The most important disadvantage of this proposal is that producer and consumer are highly coupled, which breaks one of the Linda's principles. Other studies are oriented towards solving fault tolerance questions by means of replication [12] and data dissemination [13]. These papers use the view concept in the first one and digital tuple signatures in the second one to broadcast and keep coherent the status of tuple spaces. The methods and mechanisms to develop a fault tolerant system coherent with the semantics of the original Linda model are analysed in [14]. The conclusion is that the Linda coordination language should be extended in order to support not only small systems, but also big scaled ones. Finally, the use of hash tables is used as a common starting point such as in [15,16]. An hypercube design was chosen in [15], while a hub structure arrangement was proposed in [16]. Both of them considered the communication structure and node organisation as key points. However, the proposed solution has to pay for communication overheads and the cost of keeping the information coherent among the distributed nodes.

RLinda [7] is a centralised implementation of the Linda coordination system, developed with Nets-within-Nets [6] and the Renew tool [17]. RLinda offers the Linda primitives through different protocols as RMI or SOAP. From the client's point of view, XML is used to describe tuples.In addition to technology-aware aspects, RLinda supports an extended set of matching functions, which can be used depending on the problem context and scenarios: *strong matching, weak matching, attribute matching* and the *general attribute matching* [7]. Based on these features, DRLinda is our first approach towards a distributed implementation of the Linda coordination system which focuses on data distribution and uses a centralised controller. Similarly to RLinda, the DRLinda coordination system is described in terms of Petri nets and the Renew tool and uses a pool of RLinda nodes for data distribution, thus inheriting and extending the features from RLinda. Our proposal aims at fitting into the key points for a distributed implementation. DRLinda uses the tuple concept and definition from the Renew's API and manages a mechanism to extend and change the distribution and location policies and the node configuration at run-time, which makes it an open and dynamic solution for distributed coordination scenarios. Fault tolerance, security and load balance aspects can be easily managed using customised plugins.

3   DRLinda

In this section we are going to describe the Petri net model and implementation of DRLinda. Figure 1 depicts an overview of the proposed architecture for a distributed Linda-based coordination system. The DRLinda coordinator acts as a centralised point of control in order to distribute data over the RLinda nodes. Externally, a DRLinda coordinator interface is published. Internally, it plays the client role for the set of RLinda systems used for distribution purposes. Additionally, a dynamic configuration interface is published. This interface offers the possibility of changing at run-time the node configuration (number of available RLinda nodes, location, restrictions and limits, features..) and the distribution function (based on tuple arity, on round-robin algorithms or on tuple structure, for example). Therefore, DRLinda can adapt itself to any dynamic scenario and to any dynamic node topology or distribution.

A client willing to use our DRLinda implementation may access its functionality by means of the **in**, **out** and **rd** operations via RMI or SOAP invocations, passing an XML tuple/pattern description as a parameter. For instance, the statement $t_{xml} = in(p_{xml})$ returns in t_{xml} the XML description of a tuple matching the pattern described by p_{xml}.

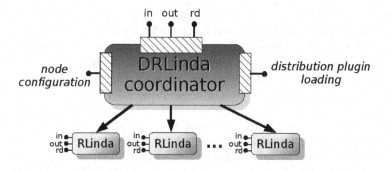

Fig. 1. The DRLinda architectural overview

The DRLinda coordinator implementation is represented in Figure 2. It shows the external interface (the client's point of view) and the internal interface (the RLinda interactions). For the sake of simplicity, the *rd* operation was hidden because of its similarity to the *in* operation. Place `available_nodes` is the node repository, where the information about the RLinda available nodes is stored and accessed in order to handle incoming operations. Place `plugins` provides a Java class or a Petri net offering distribution and allocation functions.

Let us now concentrate on the Petri net model. From a client's point of view, `in`, `rd` and `out` are invoked as single operations, just as in the original RLinda model. The `in` an `rd` operations are functions receiving a pattern as the input parameter and returning a tuple, whereas `out` is a procedure with a tuple as an input parameter. From the server's point of view, a *stub* is used to hide the sequence of operations required to perform the published operations. Renew manages *Renew-stubs*, a high-level interface description language which encapsulates an ordered sequence

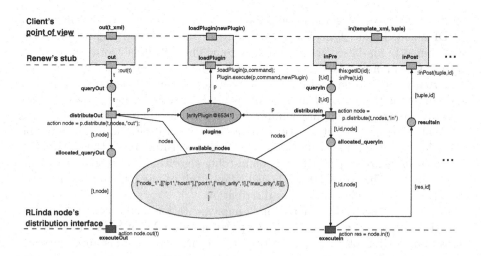

Fig. 2. The DRLinda coordination model

of transition firings. In order to do that Renew provides *channels* as a synchronous communication mechanism, making possible the transfer of parameters between the Petri net and the external generated stub-code.

Let us trace the in(p_xml) operation. From a conceptual point of view, four steps are required. First, the tuple is received from the client and passed to the Petri net through the stub code firing transition inPre. Then, firing transition distributeIn gives the node or set of nodes where a tuple corresponding to the pattern must be looked for. After that, the in operation is executed on the selected node, acting the coordinator as an RLinda client of such node (transition executeIn). Finally, the resulting tuple matching the pattern is returned to the blocked caller process, firing transition inPost. Since Renew channels are synchronous, the steps related to executing the in operation and returning the result must be separated from the server's point of view, and controlled by the Renew's stub by chaining the invocations to the corresponding synchronisation channels.

The out operation is implemented with the input channel :out(t) and results in a non-blocking operation due to the non-blocking character of the out operation in the RLinda nodes. in and rd operations result in blocking-operations as they require the use of input and output synchronisation channels. Moreover, since the behaviour of the RLinda nodes blocks a client when executing an in or rd operation, this behaviour is inherited directly by the DRLinda coordinator, which acts as a client for the RLinda nodes.

Place available_nodes provides some information about the RLinda nodes, being this information stored as Renew's tuples with the format [nodeName, hostInfo, pluginValues]. The last parameter is related to place plugins. This place contains Java objects or Renew nets offering the distribution/harvesting function distribute. The distribution function, taking a tuple and a set of RLinda nodes as parameters, gives a (non-empty) set of target RLinda nodes where the tuple can be inserted in (removed from) as a result. The harvesting function, taking a pattern and a set of RLinda nodes as parameters, gives as result a set of target RLinda nodes where the pattern can match a tuple. The plugins are modelled with a common interface, being possible to load and replace running plugins (with some reallocation and re-distribution policies) at run-time.

4   Simulation, Evaluation and Results

In order to measure DRLinda's distributed implementation performance, a benchmark, whose results are presented in this section, was executed. In the experiments a set of clients access the DRLinda coordinator. A client sets some benchmark parameters, warms-up the tuple space inserting a random number of random tuples (between 1500 and 5000) and then iterates the following sequence of operations 2000 times: 1) execute an out operation, 2) delay a random time ($T_{delay} \in [200,250]$ ms), and 3) withdraw the same tuple (operation in). When completed, the client shutdowns. A detailed justification of the used parameters can be found in [18]. To study the influence of the tuple's size we have used the same objects as in [19]: NullEntry objects (356 bytes), StringEntry objects

(503 bytes), `DoubleArrEntry` objects (1031 bytes), and `FileEntry` objects (3493 bytes approximately). The applied matching function corresponds to RLinda's *strong matching* function, as described in [7].

For the experiments, we have used a simple arity plugin, which locates a tuple in function of the number of its components. Given a tuple or a pattern, it gets the target node checking the minimum and maximum arity bounds for each RLinda node. The information of RLinda nodes was loaded into the coordinator before the benchmark was started, so that the results were obtained with no interference from management tasks.

The performance parameters chosen are: the throughput (mean number of `in`/`out` operations executed per second), and the response time (mean time between the instant in which an `in`/`out` operation arrives at the server and the time the result tuple is returned). Both parameters are measured on the server side so that the obtained results are independent of the client-server network latency.

The benchmark was executed by running q clients in the interval $[1, 60]$ in q different nodes of a cluster using the Condor software for High Throughput Computing (HTC) over 43 nodes and several additional ones dedicated to execute the DRLinda coordinator and the RLinda servers. The configuration for each node was a Genuine Intel Pentium D processor workstation, 1 GB of RAM and a SATA 7200 RPM disk subsystem running GNU/Linux with a customised 2.6.20-1 kernel compiled in order to use the `mwait` instruction and an anticipatory input-output scheduler. The Java 1.5 compiler and virtual machine were running with the default configuration, and the nodes were allowed to get the maximum available physical memory.

The chart in Figure 3(a) shows the throughput related to the number of concurrent clients q. Experiments proved that the best measures are obtained when the coordinator manages 22 RLinda nodes, so the graphical results have been limited to this configuration. The results show that the heaviest operation always corresponds to `FileEntry` objects, which require a higher CPU processing time. The best throughput are always obtained when the number of clients belongs to $[11, 15]$. Performance significantly decreases when the number of concurrent clients exceeds a certain threshold around 15 concurrent clients (this behaviour with a lower threshold was previously reported in [20,18]).

The response time is depicted in Figure 3(b). `NullEntry` objects always take less time to execute than the others. Undoubtedly, this is due to the fact that these objects are smaller than the rest and, consequently, require a lower CPU processing time. For values beyond 60 concurrent clients the CPU execution time was so high that some connections were down and the system started being degraded, making the results of no interest. Notice that in [7] results were reported up to 41 clients, while our implementation supports running up to 60 concurrent clients without degrading the system performance.

Finally, Figure 4 shows the speed-up obtained by the DRLinda implementation running the benchmark with 15 concurrent clients and varying the number of RLinda distribution nodes. DRLinda running with a single node has been

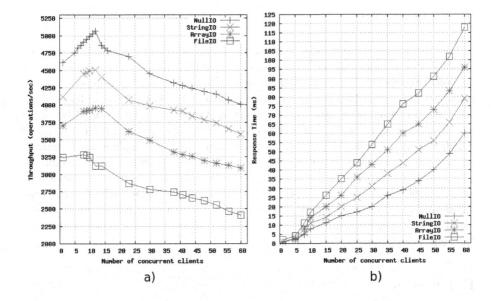

Fig. 3. Throughput and response time for the DRLinda implementation

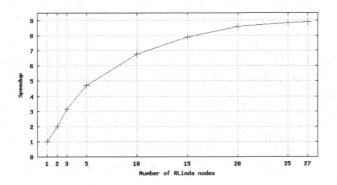

Fig. 4. Speedup of DRLinda

taken as the unity. As soon as the number of RLinda nodes grows, which means that tuples are distributed by the DRLinda coordinator over the nodes, the speedup increases in a non linear manner. For values beyond 22 deployed nodes, performance increases in so slight ratios that this value can be considered as the speed-up limit.

Comparing the execution of the benchmark using the same hardware with RLinda and with DRLinda with just one node, the results shown that DRLinda's throughput is always a 12% lower than the one obtained with RLinda. The reason of such behaviour is the extra CPU processing time and the extra latency required to process an incoming tuple or pattern in the DRLinda coordinator.

5 Conclusions

A free distributed Linda implementation should allow each business partner to interact with its own message repository. This is likely to be a common scenario in Web processes, where interactions occur in a peer-to-peer fashion. In this paper we presented DRLinda, our first approach towards a RLinda-based distributed implementation of the Linda coordination system, based on Petri Nets and Renew. Our proposal aims at data distribution with a centralised controller which uses a dynamic set of RLinda nodes and extends RLinda's functionality and features. Additionally, DRLinda can adapt itself at run-time to a specific or dynamic scenario by using dynamically loadable plugins. Therefore, we achieved an extensible implementation of the Linda coordination system, where node configuration and distribution policies can change and can be handled dynamically. Finally, the reported results shown that the distributed version clearly improves the centralised approach's performance.

References

1. Alonso, G., Casati, F., Kuno, H., Machiraju, V.: Web Services. Concepts, Architectures and Applications. Springer, Heidelberg (2004)
2. Ten-Hove, R., Walker, P.: Java Business Integration (JBI) 1.0, final release. Technical report, BEA Systems & IBM & Microsoft & SAP AG & Siebel Systems (2005)
3. Fabra, J., Alvarez, P., Bañares, J., Ezpeleta, J.: A framework for the development and execution of horizontal protocols in open BPM systems. In: Dustdar, S., Fiadeiro, J.L., Sheth, A. (eds.) BPM 2006. LNCS, vol. 4102, pp. 209–224. Springer, Heidelberg (2006)
4. Álvarez, P., Bañares, J.A., Ezpeleta, J.: Approaching Web Service Coordination and Composition by Means of Petri Nets. The Case of the Nets-Within-Nets Paradigm. In: Benatallah, B., Casati, F., Traverso, P. (eds.) ICSOC 2005. LNCS, vol. 3826, pp. 185–197. Springer, Heidelberg (2005)
5. Kummer, O.: Introduction to petri nets and reference nets. Sozionik Aktuell 1, 1–9 (2001)
6. Valk, R.: Petri Nets as Token Objects – An Introduction to Elementary Object Nets. In: Desel, J., Silva, M. (eds.) ICATPN 1998. LNCS, vol. 1420, pp. 1–25. Springer, Heidelberg (1998)
7. Fabra, J., Alvarez, P., Bañares, J.A., Ezpeleta, J.: RLinda: a Petri net based implementation of the Linda coordination paradigm for Web services interactions. In: Bauknecht, K., Pröll, B., Werthner, H. (eds.) EC-Web 2006. LNCS, vol. 4082, pp. 184–193. Springer, Heidelberg (2006)
8. Carriero, N., Gelernter, D.: Linda in context. Communications of the ACM 32, 444–458 (1989)
9. Gelernter, D.: Generative communication in linda. ACM Transactions on Programming Languages and Systems 7, 80–121 (1985)
10. Davies, N., Wade, S., Friday, A., Blair, G.: Limbo: A tuple space based platform for adaptative mobile aplications. In: Proceedings of the International Conference on Open Distributed Processing/Distributed Platforms (1997)

11. Rowstron, A., Wood, A.: An efficient distributed tuple space implementation for networks of workstations. In: Proceedings of the Second International Euro-Par Conference on Parallel Processing (1996)
12. Xu, A., Liskov, B.: A design for a fault-tolerant, distributed implementation of linda. In: Proceedings of the 19th International Symposium on Fault-Tolerance Computing (1989)
13. Patterson, L., Turner, R., Hyatt, R.: Construction of a fault tolerant distributed tuple-space. In: Proceedings of the 1993 ACM/SIGAPP symposium on Applied computing: states of the art and practice (1993)
14. Tolksdorf, R., Rowstron, A.: Evaluating fault tolerance methods for large-scale linda-like systems. In: Proceedings of the International Conference on Parallel and Distributed Processing Techniques and Applications (2000)
15. Obreiter, P., Graf, G.: Towards scalability in tuple spaces. In: Proceedings of the 2002 ACM symposium on Applied computing, ACM Press, New York (2002)
16. Bharambe, A., Agrawal, M., Seshan, S.: Mercury: Supporting scalable multiattribute range queries. In: Proceedings of SIGCOMM'04 (2002)
17. Kummer, O., Wienberg, F., Duvigneau, M., Schumacher, J., Köhler, M., Moldt, D., Rölke, H., Valk, R.: An Extensible Editor and Simulation Engine for Petri Nets: Renew. In: Cortadella, J., Reisig, W. (eds.) ICATPN 2004. LNCS, vol. 3099, pp. 484–493. Springer, Heidelberg (2004)
18. Fiedler, D., Walcott, K., Richardson, T., Kapfhammer, G.M., Amer, A., Chrysanthis, P.K.: Towards the Measurement of Tuple Space Performance. ACM SIGMETRICS Performance Evaluation Review 33, 51–62 (2005)
19. Noble, M.S., Zlateva, S.: Scientific computation with JavaSpaces. In: Hertzberger, B., Hoekstra, A.G., Williams, R. (eds.) High-Performance Computing and Networking. LNCS, vol. 2110, pp. 657–666. Springer, Heidelberg (2001)
20. Zorman, B., Kapfhammer, G.M., Roos, R.S.: Creation and analysis of a JavaSpace-based genetic algorithm. In: PDPTA '02. Proceeedings of the 8th International Conference on Parallel and Distributed Processing Techniques and Applications, vol. 3, pp. 1107–1112. CSREA Press (2002)

Object-Based Interactive Video Access for Consumer-Driven Advertising*

Guang-Ho Cha

Department of Computer Engineering, Seoul National University of Technology
Seoul 139-743, South Korea
ghcha@snut.ac.kr

Abstract. There are currently strong motivations to develop systems that can associate information with objects in video and allow users to access this information by selecting the objects in video. These systems could be used to create a new form of business applications such as consumer-driven advertising and audience-specific advertising. This paper presents the schematics for creating an object-based interactive video system and describes an authoring tool we have developed to create the interactive video. We also examine the opportunities and challenges of interactive video.

1 Introduction

Electronic advertising has become an integral earning model for many Internet-based companies but one-sided advertising is not so appealing to consumers because it could not attract the customer's interest. Advertisement banners with simple embedded images are incapable of providing consumers with a high degree of interactivity. Also, there is no merchandise browsing capability or interactive functionality implemented in these banners. Recently, rich media has elevated electronic advertising to new realms of possibility – video, audio, animation, interactive features, game and more. Rich media offers unprecedented capabilities for measuring consumer interactions with the advertisement, video plays and custom events. Interactions that rich media offer to customers improve awareness of the advertisement. However, there have been no attempts to allow users to access to the objects in the online video. If this high degree of interactivity is implemented, the general TV program and cinema themselves can become a living advertisement. The objects in the video can be associated with a specific advertisement.

In this paper, we present an object-based interactive video access model to develop consumer-driven advertising. Advertisements with entertaining components might be suitable to induce consumers to take interest in the advertisement itself. For example, we could create a TV program or movie that is an advertisement itself in which some curious viewers obtain more information about particular items such as clothes worn by an actor by selecting the objects on the monitor (or screen) with their pointing device. In other words, the TV program itself becomes an advertisement, and therefore, the viewer does

---

* This work was supported by grant No. B1220-0501-0233 from the University Fundamental Research Program of the Ministry of Information & Communication in Republic of Korea.

G. Psaila and R. Wagner (Eds.): EC-Web 2007, LNCS 4655, pp. 222–228, 2007.

not unwillingly watch the commercials embedded in TV programs. Moreover, this object-based interactive video access provides *bidirectional* connectivity between advertisers (or content authors) and consumers by inducing consumers to take interest in the objects in the electronic content.

Construction of this kind of object-based interactive video requires segmenting objects from video and providing access to the objects as part of querying and browsing the video. In this kind of video access, the viewer has a direct role in influencing the content of video. As aforementioned, a TV program or movie could be created in which viewers obtain more information such as product's brand name, price, and Web site by selecting the clothes worn by an actor during playback. Bove et al. [1] called this kind of interactive video the *hyperlinked video*. In [2, 4, 5], it is called *hypervideo*. Playback of this kind of video is well within the capabilities of today's desktop PCs. However, its creation has posed a challenge to authors because of the difficulty of identifying and tracking the selectable objects in every video frame, by either manual or automatic methods.

We have developed a "video annotator" for authoring this kind of interactive video and a "video player" for playing video and controlling the interaction between the video and the viewer. In this paper, we present the schematics for the object-based interactive video system and describe the video annotator and the video player we have developed.

In the next section, we will review the existing interactive video software tools. In Section 3, we will present our interactive video system for providing consumer-driven advertising. We conclude our paper in Section 4.

2 Related Work

Several tools have been announced for authoring hyperlinked video, most notably HyperActive [1] from MIT Media Lab and HotMedia [2] from IBM's Internet Media Group. Aimed at inserting several links into the video frame, these tools typically require a user to outline clickable regions in key frames with a bounding box or other geometrical shape, and then the tool automatically tracks the region through intermediate frames.

HyperActive pays attention to the segmentation system that classifies every pixel in every frame of a video sequence as a member of an object or group of objects. The segmentation system of HyperActive uses multiple features such as color, motion, texture, and position to improve segmentation accuracy. To break up a video into shots, HyperActive uses standard scene-change detection. In each shot, the author chooses one or more frames on which he or she defines what the desired objects are by providing the algorithm with "training data" in the form of rough scribbles on each. By using the training data, HyperActive classifies all the unlabeled pixels in the training frame and in other frames of the shot. At the end, an object map for each frame and a table are generated, and they associate each object region with some action to be performed when selected by the viewer.

Unlike the HyperActive that uses the complex method for segmenting, we employ a simple method for segmentation, that is, we describe manually the clickable objects in key frames with a minimum bounding rectangle (MBR) that encloses the selected object tightly. Advantages of using a simple geometrical description of the clickable objects include small overhead incurred in transmitting them and the ability to describe them using the standard extensible markup language (XML).

HotMedia focuses on the rich media file format that can contain heterogeneous compositions of media bit streams as well as metadata that defines the behavior, composition, and interaction semantics. Therefore, HotMedia is too heavy to use only in the interactive video access for consumer-driven advertising.

Lienhart developed techniques for annotating home movies on the fly and automatically summarizing the movies in short video abstracts [3]. His algorithms cluster time-stamped shots into semantically meaningful units, shorten shots to interesting clips, and select the clips of the video abstract and arrange them into final video abstract. His system, however, cannot incorporate real-time audience activity or other feedback during a presentation.

The object-based video retrievals presented in [6, 7] allow users to search specific video shots using the occurrence of a set of pre-defined video objects. Although these systems allow the object-based access to video, users cannot access directly the objects in the video during the video playback and the objects are pre-defined before the query time.

3   Architectural Overview of Our Interactive Video Access

The architecture of our object-based video access system is illustrated in Fig. 1. The video files with annotated metadata are archived on the server video database and are accessible to the server-side generic HTTP server. Communications for data requests, data delivery and tracking use the HTTP protocol.

On the client-side, the presentation and user interaction are controlled by our video player. The data delivered to the video player contain the raw video and the metadata containing the description for objects in the key frame in each video shot. The description for objects contains the spatio-temporal positions within the key frames and the product information for the object. Actions are triggered by user click on the clickable object.

Our video annotator architecture is illustrated in Fig. 2 and its screen capture image is shown in Fig. 3. It is a software tool for authoring interactive video that viewers can access clickable objects during the video playback. When using the video annotator, the author of the video must first digitize scenes of video. The video annotator then utilizes a simple method for *temporal segmentation* – breaking each scene up into *shots*, where a shot is used to distinguish a set of successive frames which were created by a single camera view during the shot. The video annotator calculates the differences between pixel values in temporally adjacent frames, sums them and compares them to a threshold value. If the threshold is exceeded, the two frames are likely to be on either side of a shot boundary. If not, they are probably both within the same shot. Once this process is repeated for all pairs of temporally adjacent frames, the video annotator can then identify and track the objects that appear in each shot.

In our video annotator, we manually identify the selectable objects in each shot and describe the objects with minimum bounding rectangle (MBR) that encloses the selected object tightly. Since the same objects may appear in more than one shot in

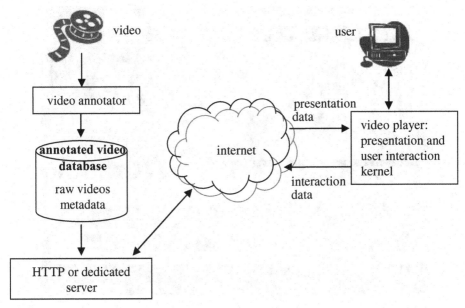

Fig. 1. System architecture

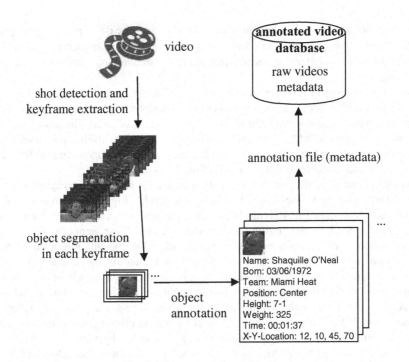

Fig. 2. Video annotator architecture

Fig. 3. Screen capture of the video annotator

most video sequences, our video annotator can associate the description for the previ-ously identified objects with the objects in a newly processed sequence images. This relieves the author's labor that spends time manually making associations between the same objects.

One must manage a number of data when authoring interactive video. To better manage data, the system creates a list of object descriptors, including semantic repre-sentations, which updates with the identification of new objects. The object descrip-tors are represented in XML. XML was chosen because of its ability to be understood by both of people and machines, its general acceptance as an open method for trans-ferring data, and its compatibility with all of the Web browsers.

On the client-side, the video presentation and user action are controlled by the video player and user interaction kernel, respectively. Fig. 4 describes the client architecture. The data file arriving at the client from the server carries annotations (metadata) as well as the multiplexed media stream. These annotations associate various actions with spatio-temporal positions within the presentation of video. Such information is controlled by the user interaction kernel on the client side. Fig 5 shows the screen capture of the video presentation and the result of user action. When viewer moves mouse on the object selectable in a video presentation, the user interaction kernel alerts the viewer as to the object is selectable by changing the mouse shape. When viewer selects the object by mouse clicking, the user interaction kernel presents the information related the object selected by viewer. In Fig. 5, the information related to the clothes worn by an actor is shown in the right panel of the video player.

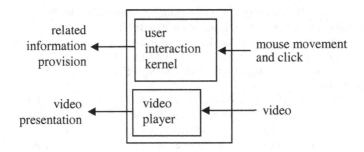

Fig. 4. Client architecture: client-side presentation and user interaction logic

Fig. 5. Screen capture of the video presentation and user action

4 Conclusion

The object-based interactive video access system presented in this paper is a toolkit for enriching e-business applications, especially, consumer-driven advertising. It provides the ability to access objects in the video during playback. It can also add hot links in the video so that viewers can jump to specific external Web sites or certain positions (i.e., shots) within the video.

We think that consumer-driven advertising is a very natural and useful advertising method because it does not interrupt viewers during watching the TV program or video. In this perspective, the object-based video access system will become a powerful tool for advertising.

References

1. Bove, V.M., Dakss, J., Chalom, E., Agamanolis, S.: Hyperlinked Television Research at the MIT Media Lab. IBM Systems Journal 39, 470–478 (2000)
2. Kumar, K.G., et al.: The HotMedia Architecture: Progressive and Interactive Rich Media for the Internet. IEEE Transactions on Multimedia 3(2), 253–267 (2001)

3. Lienhart, Rainer, Dynamic video summarization of home video. In: Proc. of SPIE 3972: Storage and Retrieval for Media Databases, pp. 378-389 (2000)
4. Pattanasri, N., Jatowt, A., Tanaka, K.: Enhancing Comprehension of Events in Video Through Explanation-on-Demand Hypervideo. In: Cham, T.-J., Cai, J., Dorai, C., Rajan, D., Chua, T.-S., Chia, L.-T. (eds.) MMM 2007. LNCS, vol. 4351, pp. 535–544. Springer, Heidelberg (2006)
5. Shipman, F., Girgensohn, A., Wilcox, L.: Generation of Interactive Multi-Level Video Summaries. In: Proc. ACM Multimedia Conference, ACM Press, New York (2003)
6. Smeaton, A.F., Jones, G.J.F., Lee, H., O'Connor, N.E., Sav, S.: Object-Based Access to TV Rushes Video. In: Lalmas, M., MacFarlane, A., Rüger, S., Tombros, A., Tsikrika, T., Yavlinsky, A. (eds.) Advances in Information Retrieval. LNCS, vol. 3936, pp. 476–479. Springer, Heidelberg (2006)
7. Sav, S., Jones, G.J.F., Lee, H., O'Connor, N.E., O'Connor, A.F.: Interactive Experiments in Object-Based Retrieval. In: Sundaram, H., Naphade, M., Smith, J.R., Rui, Y. (eds.) CIVR 2006. LNCS, vol. 4071, pp. 1–10. Springer, Heidelberg (2006)

Author Index

Lecture Notes in Computer Science

For information about Vols. 1–4575

please contact your bookseller or Springer

Vol. 4626: R.O. Weber, M.M. Richter (Eds.), Case-Based Reasoning Research and Development. XIII, 534 pages. 2007. (Sublibrary LNAI).

Vol. 4624: T. Mossakowski, U. Montanari, M. Haveraaen (Eds.), Algebra and Coalgebra in Computer Science. XI, 463 pages. 2007.

Vol. 4622: A. Menezes (Ed.), Advances in Cryptology - CRYPTO 2007. XIV, 631 pages. 2007.

Vol. 4619: F. Dehne, J.-R. Sack, N. Zeh (Eds.), Algorithms and Data Structures. XVI, 662 pages. 2007.

Vol. 4618: S.G. Akl, C.S. Calude, M.J. Dinneen, G. Rozenberg, H.T. Wareham (Eds.), Unconventional Computation. X, 243 pages. 2007.

Vol. 4617: V. Torra, Y. Narukawa, Y. Yoshida (Eds.), Modeling Decisions for Artificial Intelligence. XII, 502 pages. 2007. (Sublibrary LNAI).

Vol. 4616: A. Dress, Y. Xu, B. Zhu (Eds.), Combinatorial Optimization and Applications. XI, 390 pages. 2007.

Vol. 4615: R. de Lemos, C. Gacek, A. Romanovsky (Eds.), Architecting Dependable Systems IV. XIV, 435 pages. 2007.

Vol. 4613: F.P. Preparata, Q. Fang (Eds.), Frontiers in Algorithmics. XI, 348 pages. 2007.

Vol. 4612: I. Miguel, W. Ruml (Eds.), Abstraction, Reformulation, and Approximation. XI, 418 pages. 2007. (Sublibrary LNAI).

Vol. 4611: J. Indulska, J. Ma, L.T. Yang, T. Ungerer, J. Cao (Eds.), Ubiquitous Intelligence and Computing. XXIII, 1257 pages. 2007.

Vol. 4610: B. Xiao, L.T. Yang, J. Ma, C. Muller-Schloer, Y. Hua (Eds.), Autonomic and Trusted Computing. XVIII, 571 pages. 2007.

Vol. 4609: E. Ernst (Ed.), ECOOP 2007 – Object-Oriented Programming. XIII, 625 pages. 2007.

Vol. 4608: H.W. Schmidt, I. Crnkovic, G.T. Heineman, J.A. Stafford (Eds.), Component-Based Software Engineering. XII, 283 pages. 2007.

Vol. 4607: L. Baresi, P. Fraternali, G.-J. Houben (Eds.), Web Engineering. XVI, 576 pages. 2007.

Vol. 4606: A. Pras, M. van Sinderen (Eds.), Dependable and Adaptable Networks and Services. XIV, 149 pages. 2007.

Vol. 4605: D. Papadias, D. Zhang, G. Kollios (Eds.), Advances in Spatial and Temporal Databases. X, 479 pages. 2007.

Vol. 4604: U. Priss, S. Polovina, R. Hill (Eds.), Conceptual Structures: Knowledge Architectures for Smart Applications. XII, 514 pages. 2007. (Sublibrary LNAI).

Vol. 4603: F. Pfenning (Ed.), Automated Deduction – CADE-21. XII, 522 pages. 2007. (Sublibrary LNAI).

Vol. 4602: S. Barker, G.-J. Ahn (Eds.), Data and Applications Security XXI. X, 291 pages. 2007.

Vol. 4600: H. Comon-Lundh, C. Kirchner, H. Kirchner (Eds.), Rewriting, Computation and Proof. XVI, 273 pages. 2007.

Vol. 4599: S. Vassiliadis, M. Berekovic, T.D. Hämäläinen (Eds.), Embedded Computer Systems: Architectures, Modeling, and Simulation. XVIII, 466 pages. 2007.

Vol. 4598: G. Lin (Ed.), Computing and Combinatorics. XII, 570 pages. 2007.

Vol. 4597: P. Perner (Ed.), Advances in Data Mining. XI, 353 pages. 2007. (Sublibrary LNAI).

Vol. 4596: L. Arge, C. Cachin, T. Jurdziński, A. Tarlecki (Eds.), Automata, Languages and Programming. XVII, 953 pages. 2007.

Vol. 4595: D. Bošnački, S. Edelkamp (Eds.), Model Checking Software. X, 285 pages. 2007.

Vol. 4594: R. Bellazzi, A. Abu-Hanna, J. Hunter (Eds.), Artificial Intelligence in Medicine. XVI, 509 pages. 2007. (Sublibrary LNAI).

Vol. 4593: A. Biryukov (Ed.), Fast Software Encryption. XI, 467 pages. 2007.

Vol. 4592: Z. Kedad, N. Lammari, E. Métais, F. Meziane, Y. Rezgui (Eds.), Natural Language Processing and Information Systems. XIV, 442 pages. 2007.

Vol. 4591: J. Davies, J. Gibbons (Eds.), Integrated Formal Methods. IX, 660 pages. 2007.

Vol. 4590: W. Damm, H. Hermanns (Eds.), Computer Aided Verification. XV, 562 pages. 2007.

Vol. 4589: J. Münch, P. Abrahamsson (Eds.), Product-Focused Software Process Improvement. XII, 414 pages. 2007.

Vol. 4588: T. Harju, J. Karhumäki, A. Lepistö (Eds.), Developments in Language Theory. XI, 423 pages. 2007.

Vol. 4587: R. Cooper, J. Kennedy (Eds.), Data Management. XIII, 259 pages. 2007.

Vol. 4586: J. Pieprzyk, H. Ghodosi, E. Dawson (Eds.), Information Security and Privacy. XIV, 476 pages. 2007.

Vol. 4585: M. Kryszkiewicz, J.F. Peters, H. Rybinski, A. Skowron (Eds.), Rough Sets and Intelligent Systems Paradigms. XIX, 836 pages. 2007. (Sublibrary LNAI).

Vol. 4584: N. Karssemeijer, B. Lelieveldt (Eds.), Information Processing in Medical Imaging. XX, 777 pages. 2007.

Vol. 4583: S.R. Della Rocca (Ed.), Typed Lambda Calculi and Applications. X, 397 pages. 2007.

Vol. 4582: J. Lopez, P. Samarati, J.L. Ferrer (Eds.), Public Key Infrastructure. XI, 375 pages. 2007.

Vol. 4581: A. Petrenko, M. Veanes, J. Tretmans, W. Grieskamp (Eds.), Testing of Software and Communicating Systems. XII, 379 pages. 2007.

Vol. 4580: B. Ma, K. Zhang (Eds.), Combinatorial Pattern Matching. XII, 366 pages. 2007.

Vol. 4579: B. M. Hämmerli, R. Sommer (Eds.), Detection of Intrusions and Malware, and Vulnerability Assessment. X, 251 pages. 2007.

Vol. 4578: F. Masulli, S. Mitra, G. Pasi (Eds.), Applications of Fuzzy Sets Theory. XVIII, 693 pages. 2007. (Sublibrary LNAI).

Vol. 4577: N. Sebe, Y. Liu, Y.-t. Zhuang, T.S. Huang (Eds.), Multimedia Content Analysis and Mining. XIII, 513 pages. 2007.

Vol. 4576: D. Leivant, R. de Queiroz (Eds.), Logic, Language, Information and Computation. X, 363 pages. 2007.